HIGH PRESSURE WATER JETTING

AN OPERATOR'S MANUAL

TIM EVEREST

This book is copyright and all rights are reserved
by Tim Everest © 2012

Reproduction in any manner whatsoever without the written permission of Tim Everest is strictly forbidden.

Manufacturer and supplier trademarks and/or trade names have been used for clarity or reference in this document to refer to either the entities claiming the marks and names of their products. Tim Everest disclaims any proprietary interest in trademarks or trade names herein.

Perth Western Australia April 2012 Rev: 1.1

ISBN: 1479183121
ISBN-13: 9781479183128
Library of Congress Control Number: 2012915951
CreateSpace Independent Publishing Platform
North Charleston, South Carolina

CONTENTS

Chapter One – Legislation 1
Chapter Two – Hazard Identification and Management 7
Chapter Three – Being a Professional 29
Chapter Four – Understanding and
checking your equipment 43
Chapter Five – Understanding Drains and Pipe 87
Chapter Six – What nozzle holder to use?. 111
Chapter Seven – Nozzles and nozzle holders 117
Chapter Eight – Nozzle bore selection and Friction Loss . . 143
Chapter Nine – Designing a nozzle holder 153
Chapter Ten – Fittings, Threads and Joints. 163
Chapter Eleven – Caring for your Hose. 179
Chapter Twelve – Tools you can make, gun use and tips . . 191
Chapter Thirteen – Confined Space Entry. 207
Chapter Fourteen – HP Water Jet Injuries and 1st Aid. 271
Chapter Fifteen – Personal Protective Equipment 277
Chapter Sixteen – Vacuum Loading 283
Useful Numbers. 297
Drain Cleaning Simplified Sketches 302

FAST FIND CONTENTS

Introduction ix
Some Standards that may apply 1
Hazard Identification and treatment 7
Hazard Rating Chart 11
HAZID break down chart example 14
HAZID Checklist example 18
Work Plan example 21
Typical Site Safety Prestart Checklist example 23
Professionalism and Public Relations 29
Other Customers and the public 33
Understanding and checking equipment 43
The pump and how it works 45
Slippage 56
Cavitation 58
The Engine 61
The Coupling 65
The clutch 65
Diesel Fuel 67
Radiator and Thermostat 69
Air Cleaner 71
Batteries 73
Fan Belts 76
Turbochargers 78
Cold Starts 81
Daily Maintenance Checklist example 83
Daily Equipment Checklist Example 84
Understanding Pipes and Drains 87
Nozzle Bore Selection & Friction Loss 143

Nozzle Flow Chart 145
Hose Friction Chart 147
Designing a pipe cleaning nozzle holder (Bomb) 153
Fittings Seals and Joints 163
Threads Identified 164
Caring for your hoses 179
Some tooling to make—water lifts
& Air Lifts. Tips and thoughts. 191
Using a hand jetting gun. 198
Confined Space Entry 207
Water Jet Injuries & First Aid. 271
Personal Protective Equipment (PPE) 277
Vacuum Loading 283
Tim's Numbers. 297
How to clean a pipe—pictures 302

Acknowledgments

I would like to thank my long-suffering wife, Maeve, whose friendly nagging kept me at it, and Danek Liwszyk of Jetcut Pty. Ltd., for his encouragement and helping me think outside the square. I also thank Sajith (Sarge) Jayamaha for being the sounding board for some of my crazy ideas. Finally to all those many operators I have trained over the years who asked the questions. Thanks to you all.

Introduction

This book has been written to assist water jet operators working on surface cleaning, concrete rehabilitation, surface preparation, and clearing pipes, drains, and sewers. It introduces a little of the theory and practice gained by the author over forty years working in the industry. I started off operating water jets underwater in the North Sea as a diver, became really interested in water jet operation, and moved my interests from under water to all aspects of water jetting from the little workshop units right up to 4000 bar (60,000 psi), million-plus-dollar machines. I have physically worked in just about every water jet application known. Later in life, I decided to go into business in Perth Western Australia to make my own tooling. My company was called T Squared Developments, and I took on the difficult things and made tooling for the application.

As the reader should know, nothing beats hands-on experience, and I cannot include every trick I have discovered over the years into a small book. I provide some guidelines and hints and tips to get you started, but you will have to take it from here on and develop your skills.

The number-one requirement is to *experiment*. If you have an idea, try it out and see if it works. If it does work, *document it*. If you do not write it down, maybe by this time next week you will have forgotten what you did that was so successful and "earth shattering". You know the saying, "The only guy who never makes a cock up is the guy who will not take a chance and try". *It's true.*

High Pressure Water Jetting – An Operator's Manual

You might consider also documenting the things that did not work, so you do not repeat them later. Write these in a good-quality, hard-cover notebook, and keep it safely. This notebook should be your number-one tool; you will refer to it for years. Do not be concerned about your spelling or handwriting if that may hold you back. This will be *your* notebook, and no one else will see it. Make sketches, scribble notes, and record flows and pressures used. Do not forget to put the date on the top of each page for easy reference. A regularly kept notebook can be a good reference tool in the event of an accident, incident, or query regarding a job you did some time ago.

I do not dwell on ultra-high pressure, as I feel that is a specialist process and not truly a tool for everyday water jetting operations. This book deals, in the main, with pressures up to about 1400 bar, or 20,000 psi, which is the optimum range for day-to-day work. I will prepare a book on ultra-high next, hopefully working with Danek from Jetcut in Perth.

All instructions, suggestions, and procedures have been used or are in use by others in various parts of the world. Where applicable, the actions and instructions are based on the requirements of various standards and codes of practice currently in force. Recommendations and suggestions contained herein are provided to assist you in doing your job in an efficient manner. It is, however, expected that you apply common sense to your actions and not expose yourself to hazards. The author does not accept any responsibility nor is responsibility implied in the event of an accident as a result of any recommended activity.

I wanted to include photos of various items of pumps and equipment from some of the major manufacturers. Despite several requests for permission to do so, I received no reply from any of them. Any equipment photos in this manual are items manufactured by me.

Introduction

Above all, have fun doing your job. Water jetting is fun, and once you have a few wins on the board, you are on your way. If it is not fun, it is time to change your vocation. Oh and in case I forget, wear your PPE, I now have two hearing aids costing about the same price as a small car, because I was too macho to wear ear muffs or plugs!

There are many women working in the water jetting industry today. Throughout this book references to "he" or "him" includes "she" or "her". I apologise ladies if you feel I am ignoring you. I am not, one of the best operators I know is a lady!

Chapter One

Legislation

SOME STANDARDS AND CODES THAT MAY APPLY
I list only those applicable to Australia and New Zealand. Readers in other countries should obtain relevant local standards and codes of practice. Any accident or injury sustained for which you may be prosecuted under the Health & Safety in Employment (HASE) Act 1993 may be as a result of a contravention of one or another Australian or New Zealand Act, Standard (AS/NZ XXXX), or Code of Practice. It is advisable to be aware of acts, regulations, standards and code requirements. *Acts* and *regulations* are law and must be complied with, standards and codes are guidance notes.

There are a few areas where no Australian standard is available. Under these circumstances, US, DNV, DIN, or British (BS) may apply. Ignorance of an applicable industry accepted standard is no excuse.

AS/NZS 4233.1:1999 HP Water Jet Systems (Guidelines for Safe Operation & Maintenance). If your country does not have a standard, get a copy of these two documents from the STANDARDS AUSTRALIA website.

AS/NZS 4233.2:1999 HP Water Jet Systems (Construction and performance)

AS/NZS 1270:2002 Hearing protection

AS1319:1994 Safety signs for the occupational environment

AS 1742 Manual of uniform traffic control devices

AS 1742.3:2002 Manual of uniform traffic control devices—Traffic control devices for works on roads

AS 2865:2001 Safe working in a confined space

AS 3765 (plus parts 1 and 2) Clothing for protection against hazardous chemicals See also AS/NZS 4501.2 2006

AS/NZS 4501.2.2006 Occupational Protective Clothing—General Requirements

AS/NZS 1337:1992 Eye protectors for industrial applications (plus Amdt 1:1994 and Amdt 2:1997)

AS/NZS 1715:1994 Selection, use and maintenance of respiratory protective devices

AS/NZS 1801:1997 (plus Amdt 1:1999) Occupational protective helmets

AS/NZS 1891 1995 Industrial fall-arrest systems and devices—Safety belts and harnesses (plus amendments 1-5)

AS/NZS 1891.3:1997 Industrial fall-arrest systems and devices—Fall-arrest devices

AS/NZS 2161.1:2000 Occupational protective gloves—Selection, use, and maintenance

AS/NZS 2161.2:2005 Occupational protective gloves—General requirements

Legislation

AS/NZS 2161.3:2005 Occupational protective gloves—Protection against mechanical risks

AS/NZS 2161.5:1998 Occupational protective gloves—Protection against cold

AS/NZS 2210.1:1994 Occupational protective footwear—Guide to selection, care, and use

AS/NZS 2210.2:2000 Occupational protective footwear—Requirements and test methods (plus amendments 1-3)

AS/NZS 3191:2003 Cords electric, flexible

AS/NZS 3160:2001 Approval and test specification hand-held, portable electric tools (plus amendments 1 and 2)

AS 1657:1992 Fixed platforms, walkways, stairways, and ladders—Design, construction, and installation

Other reading:

OH&S, Accident reporting and investigation

OH&S Safety Supervisors Guide 1987 (Safety in Construction No. 19)

The Construction Act 1959, Scaffolding.

OH&S Scaffolding, Minimum Standards 1991

OH&S Cranes & Lifting Appliances 1985 (Safety in Construction No. 24)

That's a lot of reading.

You are not expected to know them all, but you might consider looking up the act, regulation, standard, or code of practice when a query arises. Someone has addressed every action you take at work; it is either somewhere in a standard, regulation (in support of a standard), or covered by a code of practice. You *must* read AS/NZS 4233 1999, parts 1 and 2; these apply specifically to water jetting.

Typically you cannot be prosecuted for not adhering to a standard; standards are normally advisory documents and published as "the best available information". However, in the event that an injury is taken to court, the judge or magistrate may use the standard as the "Current Industry State of Knowledge" and club you with it. If working on an oil field installation and the standard is referred to in the Safety Case, then it does become a legal document and has all the teeth of an act. Again, acts are *law* and must be obeyed, as must the regulations supporting those acts.

If you have a situation that worries you, select a standard from the list above and ask your employer to get you a copy. Larger companies are members of Standards Australia and can simply pull them off the Internet. Occupational Health and Safety Inspectorate (Work Safe, Work Cover, etc.) supplies most of its data for free. Pal up to your local OH&S representative; he or she can be a very useful friend. OH&S's function is to ensure *your* safety at work. It is there to help and should not be seen as the enemy or Big Brother. Your input and questions are appreciated, as they help OH&S to help others.

Always play the "what if" game. What will happen if that goes wrong? Will someone get hurt? Will equipment get damaged? Will we cause pollution, noise, or obstruction? Know the hazards. Ask the questions, get the answers, and manage the hazards before they occur. Water jetting is your profession; a true professional knows the limitations of his job both in aspects of safety and in capability. The only place you will get that

Legislation

information is by talking about it, reading it, and remembering what you have read or heard.

And finally, do obey the rules. They are there to ensure that you get home safely and in one piece tonight. It may take only a few minutes of your day to check for the hazards. Think—is this a safe place to work? If it is not, make it so, or do not start. You have a legal right not to undertake a job you think is too dangerous.

Above all, do not be afraid to ask. If you have a question, you can be sure others have that same question. They too would like an answer or a query sorted out. It's a good feeling being able to help. All the major pump manufacturers have internet chat rooms—use them. There are others out there who need your help, or you might need theirs, so talk to them. If all else fails contact the author, Tim Everest, in Western Australia:

Ph: +61 (8) 9206 3172

Mob: 0437 202 576

E-mail: teverest11@bigpond.com.

If there is an answer, I can find it.

Chapter Two

Hazard Identification and Treatment

WHAT IS A *HAZARD*?
A hazard is an identifiable indication of an event that may injure you, other people, your equipment, or the environment. It is not necessarily something that has already happened (this is normally already covered by a near-miss report or similar and dealt with in a JSA [Job Safety Analysis], JHA, or JRA). But it is something you identify that could happen during or as a result of your work. Do not depend on the JSA as the answer to all things. Situations and sites change—sometimes overnight. Be observant, seek out hazards, and address them.

Potential injuries to self are easily recognizable—they hurt. Injuries to plant and equipment are often ignored. Most of these are caused by carelessness and a disregard for the cost and ongoing safety of the equipment we use. An injury or near miss to plant is equally deserving of an accident report and an investigation. Why did it happen? How did it happen? What are we going to do about it? If you ask "Why?" three times, you *will* get an answer you can act on.

Injuries to the environment are becoming more and more frequent. People often think that their little bit of pollution does not really matter, but together they may mount up to a major environmental catastrophe. We must look after our environment so that we leave something good for the next generation. Report, document, and investigate injuries to the environment.

Acid running into a storm water drain will eventually end up in a river or the sea, killing fish and destroying habitat. Be aware of your pollution. If caught polluting, *you* are responsible, not the client or your employer. *You* will go to jail, not them. If in doubt about runoff, *ask somebody* and pass the responsibility along, but make sure you get your instruction in writing. If you get no satisfaction, ask the Department of the Environment or Occupational Health and Safety (OH&S). That is their job.

You are not necessarily the cause of a hazard; it could be the result of another person's or normal site activity that may affect your ability to do your job safely. What is a hazard to you may be the normal activity of someone else.

For example, when steam pressure increases beyond a set point, the control-room operator lets off some steam to reduce the pressure. Normally, this procedure is an accepted, common practice. However, should you be asked to descale or clean a pipe close to the outlet or discharge point, you could be scalded to death by the steam.

Hazard management procedures would typically be to isolate the area with barricade tape and/or to inform the operator that you are there so he can clear the area before he relieves pressure. A written, step-by-step procedure is established, and the hazard is managed. The management plan may be simple or complex, depending on the complexity of the operation.

There are three possible steps to manage a hazard:

Hazard Identification and Treatment

1. To **eliminate** it completely, shut off the boiler, or do not do the job.

2. To **isolate** the worker from the risk of injury, use procedures, guards, and barriers and wear of personal protective clothing

3. To **minimise** the possibility of harm, the job has to be done and steam cannot be shut off. We establish a management procedure that is documented and tested to ensure that it works and minimises the risk of injury.

Hazards need to be graded to make sure we have clearly identified the outcome of an incident. Is it likely or unlikely to happen, or is there a vague chance that it might happen?

Hazards need to be classified into the likely outcome of the identified hazard. Will it kill one or more persons? Will it seriously injure one or more persons? Will it cause a minor injury requiring a quick trip to first aid?

The purpose of grading a hazard is to establish clearly in one's mind the amount of risk involved. Armed with this data, we can then address the hazard to see whether *eliminate, isolate,* or *minimise* should or could be applied to it to reduce its risk level. Proper management and common sense (not so commonly found as one might think) can *always* reduce the likelihood and/or the severity of an accident.

In some cases the hazard cannot be addressed and the job cannot be done. This is where that word *practicable* comes into all standards, acts, and codes of practice. We need to make the job as safe as is practicable. But how we arrive at practicable is a hard one. There is a limit to the amount of safety features you can put in place; a cost of a quarter-million dollars for safety to do a fifty-dollar job is not practicable. So, the acts add the word *reasonable*. Is it practicable to prevent an incident/accident; is it reasonable to expect that you be protected from it?

I have a tow bar on my car that sticks out behind the car about 300 mm, I walked into it and did some nasty damage to my shin. So I looked around, and for twelve dollars, I got a purpose-designed rubber cover that protects anyone. It was practicable (very cheap), and it was reasonable to expect others to do the same. The Duty of Care Act states that we must take all reasonable and practicable care to ensure the safety of others. I identified a hazard for which I was responsible, and I found a solution that would *minimise* the risk to others.

Hazard management plans do not involve you doing a Hazard Management Sheet by yourself; you can involve several groups, departments, or activity centres. Rope them in; they are all a part of the safety function:

- Project planning—for example, when the job happens, the boilers are shut down for maintenance

- Engineering—modification of tooling or access, robots?

- Administration—training, supervision and monitoring

- Safety—Personal Protection Equipment (PPE) issue and training—the provision of PPE to protect the worker from the identified hazard. PPE is the very last resort and should be avoided if at all possible. PPE does not indicate that the operator is safe and that you have addressed the hazard adequately. What happens if he does not wear it (a common activity), it is incorrectly fitted, or it is inadequate for this particular application?

Hazard Identification and Treatment

		Fatality, one or more	Permanent Disability	Lost time injury	Medical treatment	First Aid no lost time
		1	2	3	4	5
Will happen	1	1	2	3	4	5
Common/ likely	2	2	4	6	8	10
May happen	3	3	6	9	12	15
Rarely happens	4	4	8	12	16	20
Unlikely to happen	5	5	10	15	20	25

Figure 2.1 Hazard Weighting Chart

The simplified Hazard Weighting Chart is typical of the tools safety managements use. The likelihood of the event happening is across the top, with 1 being most likely, and the injury resulting is down the side, with 1 killing one or several persons.

Use your judgment in deciding the severity and likelihood. It is always smart to think pessimistically when deciding the level. We find ways of reducing the number as we manage the hazard. Your Hazard Weighting Chart will show the result without the management and then show the result after management.

Be aware that the hazard analysis could result in you not doing the job at all. If you cannot get the rating down, talk to your safety officer or the client safety representative on site. Once you identify a hazard, they must help you to reduce its impact by law. Always look for the Worst Credible Outcome, or WCO. If we tripped today but managed to grab a rail and caused a minor sprain, could the WCO be broken bones from a nasty fall?

The word *credible* needs a little translation: what could we *reasonably* expect to have happened if we had not caught hold of the rail?

Let us identify how some situations might be managed.

1. *Clear a pipe next to the steam relief point.*

If we do not address the problem what could be the result? A fatality (1) will happen or (2) or is likely to happen.

The object of our analysis is to get the number down into the shaded areas 12 to 25 by multiplying the numbers. The ideal outcome is the total elimination of any hazard. Some hazards simply cannot be removed as the job must be done.

So, back to the pipe clearing job next to the steam outlet point: We may discuss setting up a procedure where the control room operator was requested to inform us by radio link that he was going to blow off steam. If we put this procedure in place, what is the situation? He could forget. We might not hear his warning. One of our crew might move in the wrong direction—into danger rather than out of it. So what is the number now?

A fatality (1) may happen (3). (1 X 3=3) is unfortunately far too high.

So let us look at a procedure where the operator informs us. What if he informs us, we clear the area, we inform him that the area is clear, and then he blows off steam. As above, he could forget that we are there.

Hazard Identification and Treatment

So, that does not work. What if we put a DANGER tag on his valve handle? That might work, but he has been doing this job for the last five years, venting several times per shift, so it is not likely. But any distraction and he may well overlook the tag.

A fatality (1) rarely happens (4). This is still too high.

So, let us look at engineering the hazard out. Could we fit a bend onto his pipe to direct it away from us. That might be possible, but this may require administration involvement and a delay. The customer is in a hurry. We cannot get away from a serious accident.

It is starting to look as if we cannot manage this hazard today. So now what do we do? We decline the job due to the hazard. We tell the client he needs to shut down the boiler and isolate the circuit while we do our job.

To be sure of the isolation, we check that the valves are closed, and we tag the valve with our personal DANGER tags. Presto, *no serious life-threatening hazard.*

However, even though the blow-off valve is closed and tagged, the pipe is still very hot. If we lean against it, we will be burnt. Let us look at this. We could put a barricade next to the pipe to inform all that there is an existing hazard (HAZIDs). Now the numbers look like this:

It rarely happens (4). But if it did it, it would most likely require medical treatment (4), which would come to 16.

This is now an acceptable risk, one that can be monitored by supervision and minimised or eliminated by instruction (training). We need to get the risk down As Low As Reasonably Possible, or ALARP

High Pressure Water Jetting – An Operator's Manual

This is a simple procedure to follow, but a valuable one. Using the numbers *makes* you think about the potential danger of a job and *makes* you try to reduce the risk and outcome to ALARP.

2. Bend down to pick up a heavy hose.

An injury may happen (3), causing a lost-time injury (3), which adds up to 9, which is not acceptable. *Get some help.*

Job assessment: Prior to any job, we need to a have a short meeting, normally called a toolbox meeting. All crew working on the job plus any other interested parties who may contribute information attends this meeting. This meeting is used to identify the job steps that need to be taken, identify the hazards associated with those steps, and prepare the documentation:

Work Instruction: Hazards Identification and Management (HAZIDs)

Hazard identification procedures indicate what is to be done and the steps to be taken to do it. If you do not have a HAZ-IDs for loading the truck, you must do one of those too. I will assume you do. I detail below what steps are involved.

HAZID? Has a hazard been identified doing this step?

Activity	Action	HAZID?
Drive to site	John Smith driving—long trip—alternate licensed driver? Safe vehicle?—pre start checks carried out?	Yes
Arrive on site	Entry to site—all unfamiliarity?	Yes

Hazard Identification and Treatment

Set up on site	Establish workplace—carrying heavy equipment	Yes
	Can we use forklift?	Yes
	Is prestart housekeeping necessary?	Yes
	Who will drive forklift? Peter has a ticket	No
	Set up barriers	Yes
	Set up foot valve and observer station	Yes
	Flush hose system clean	No
	Join hoses and test	Yes
	Select and fit nozzle holder	Yes
	Insert holder in pipe	Yes
	Leaks and hose burst/blow off	Yes

Carry out the work	Clear blocked pipe	
	Contents of pipe splash back to oppo	Yes
	Spoil management	Yes
	Water runoff	Yes
	Access—scaffold	Yes
	Noise	Yes
	Strains and sprains	Yes
	Slips and falls	Yes
Break down site	Remove equipment and pack up truck	
	Lifting heavy hoses	Yes
	Slippery underfoot	Yes
	Housekeeping and tidy site	Yes
Leave the site	Driving on site	Yes
Drive back to base	Long drive—tired crew	Yes

Figure 2.2. STEPS breakdown

Now that you have broken the job down into little segments, you can simply pick out the hazards and deal with them easily.

Hazard Identification and Treatment

The form above will assist you in deciding what to do, who does it, and how you check the result. The "who does it" bit is most important. We need someone to climb up a ladder on the outside of a tank to carry a rope up. Joe Jones will do it. Have we asked Joe? No. When we get on site, everyone gets on with their job. But Joe is afraid of heights, so he's a quivering mess hiding behind a heat exchanger around the side of the tank. And Fred will drive the cherry picker. Does Fred have an Elevated Work Platform Ticket? No. Do not simply arrive on site and issue instructions. You *will* come unglued.

When deciding the *how* bit, detail the job: Hose from A to B—how far is it? Forty-five metres. That's three 20 metre hoses, two joiners, suitable spanners, a tin of anti seize paste, six hose stockings. Does the tank ladder have security hoops/back scratcher? Can the climber fall off? Does he need a harness and sliding fall arrestor? *Asking the questions and getting the answers could save a life and lots of time messing about before the job can start.*

Now *check* that it works` or happens. How do we do that? Who checks? What reminds him or her to check? We make a checklist. How? We refer to our training manuals; we refer to previous work instructions; we build it at the toolbox meeting; we delegate someone to check, or the supervisor takes it upon himself to check using a checklist.

Do not rely on your memory only, you *will* forget. When the client is bitching and the rain is running down your neck, it is all too much trouble to remember. A checklist helps, believe me.

When identifying hazards, please think carefully about the possibility of something going wrong. It will, with regular monotony.

Driving to a site with a heavy vehicle *is* a hazardous occupation. Think about it: there are wreckers yards crammed with smashed trucks to prove that this simple (not so simple) procedure—driving to a site—is hazardous and must be addressed in the planning stage. Driving is a common thing, and we all do it—some better than others. Some have a point to prove; others simply have a heavy foot. And most of us think, *Simple, I can do it my sleep.* Yeah, that's the problem.

Also, it might not be you, you have no control over the other morons on the road and the way they drive or the condition of their brakes, tyres, steering, lights, mirrors, or load. They are out to get *you*. In my opinion, the most hazardous thing you will do today will be to drive your truck on the open road. On site, *you* control the hazards; on the road, your guardian angel does. Make sure you do not drive faster than she can fly.

It follows that driving to a site is the first item on the list. Driving home again should be the last. Which is the most hazardous? Driving home. You are wet, tired, and stiff; the sun is going down; and you are most likely driving into the sunset. As sure as nuts, your windscreen is covered in soil spatter from the job. Can you see the road ahead properly? Are your socks wet and your boots full of water? You have a bunch of potential hazards that need to be addressed and acted on.

Activity and possible hazard	Rating	Eliminate	Isolate	Minimise	How?	Who?	Rating	Check
6 hour drive to site—tired driver = fatality	3			£	Regular rest periods and rotate driver	Peter and John	5	Monitor each other
Entry to site—unfamiliar with site and activity taking place—unexpected incident causing LTA	9	£			All crew undertake site induction. Ask client about hazards	All crew	25	Inducted. Checklist

Hazard Identification and Treatment

Site set up—access restricted, heavy equipment, working off scaffold—strains slips and falls.	9	£	Inspect site, check Skaftag, and boards, secure ladder. Get help to move heavy items—forklift?	All crew	16	Help and monitor each other. Checklist
System leak test—leak impact onto crew or other trades, hose burst or blow off fitting—penetrating wound	9	£	Barricade tape and crew paying attention. Fit restrainers to hose joints	All crew	25	Supervisor monitor. Checklist
Contents of pipe are acidic; splash injury causing serious chemical burns	4	£	Fit exit splashguard, wear PPE, gloves, monos and face shields. Check operation of eye wash and showers	All crew	12	Charge hand and checklist

Figure 2.3. Hazard management checklist

You have just put in ten hours of hard graft, and now you are planning a four-hour drive home. Should you not check into a motel and have a sleep and start fresh in the morning? When you get into the cab, think, *Is this a safe place to work right now?* If not, *do something about it*. The cost of putting you and your crew up for one night in a motel is negligible compared to a wrecked truck and three guys in hospital.

As you can see, it is not always possible to reduce or eliminate the hazard to an acceptable level. Driving to a site is always a very hazardous occupation that we take for granted. This is an area where the word *practicable* comes into play; you must make the job as safe as is practicable. If you come up against a 5.6

or 7 that you cannot reduce, then I would recommend that you decline the job until the client can help you put safety features in place that do reduce the risk. Under the Occupational Health and Safety Act, your employer and/or client must, as far as is reasonably practicable, provide you with a safe place of work. Make it so.

To prevent the possibility of splashback out of the pipe, using a Rotofan and set yourself up one or more hose lengths away from the pipe entry point. That may reduce the rating from 6 (not the best) to as little as 15 (excellent).

So, you can now see why we do these hazard Identification procedures and how they can be a very valuable tool. Another name for this procedure is a Job Risk Analysis, or Job Safety Analysis (JSA). You need to think carefully about the job you are about to do and to use your imagination to identify the worst possible thing happening, or WCO. (It will unless you manage it.). This will not only help you manage the hazard, but also will inform you of areas where extra care needs to be taken.

Once the hazard identifications (HAZIDs) have been done, you can prepare a work plan or instruction. The outcome of the HAZIDs, normally a simple two-column list, will detail the safety equipment and manpower you need for each aspect of the job.

The work plan will give a step-by-step instruction of how you plan to do the job and who does what and when. Although safety is paramount, this document relates to step-by-step actions of individuals primarily.

Other things normally in the work plan or instruction include the following:

- Site address

- Client name and contact

- Emergency contact

Hazard Identification and Treatment

- Title of the equipment you are to work on

- Details of crew

- Details of equipment to be carried on the truck

- Inductions required

- Details of environmental management (This should also be addressed in your HAZIDs document and include where your spoil is going, who cleans up, preparatory work you need to do before you can start cleaning, and so on.)

Armed with a proper plan, you will do the job better, safer, and sooner, and you will be on your way home without mishap.

WORK PLAN (Example)

JOB		CLIENT	
Descale acid-waste drain pipe		*Gasblowers Limited*	
adjacent to steam blow off valve area		*62—68 Another Street*	
16—Steam control room		*Colombo South*	
Crew: *Peter (Supervisor), John, Paul*		*Ph. John Smith (06) 334 4567*	
Pump 27, see attached equip. list		*Site Emergency 4444*	
Environmental HAZID: *Waste product/used water is contaminated with dilute hydrochloric acid; waste must not go to storm water drains, bund area and seal local storm water outlet (Client action). Dispose via vacuum truck (HAZCHEM permit?) to approved discharge point.*			
PPE requirements: *All persons inside barricade area to wear mono goggles, rubber gauntlets, acid-proof wets, steel-toed rubber boots. Check location and function of shower and eye wash station.*			
STEP		HAZ	CHECK
Travel to site		5	*Change driver hourly Peter/John*
Arrival on site, site inductions		25	*Induct all/client*
Carry out isolations and sign on in control room		25	*Check all signed in, check isolations /supervisor*
Set up site:			
Site hazards, eyes, ears, feet		16	*Issue PPE, wear PPE. See HAZIDs*

Housekeeping prior to job start	16	Get help to move scaffold stack Inspect access and egress to area Prestart checklist/ all
Waste management (environmental)	10	Check bunding, catchment sump, and storm-water drain seals. Prestart checklist, HAZIDs/ Peter
Set up barricade tape and signs; check above and below workplace. Barricade/baffle exit pipe end	16	Visually alert for site activities, walk around. Vac-truck running and checked? Prestart checks done? Peter
Flush and test hoses and install with restrainers as required after full inspection of hose and fittings	25	See hose checklist/John
Carry out prestart engine and pump checks	12	See checklists /Paul
Check workstation, inspect scaffold and Scaftags.	18	Visually inspect all crew
Fit starter bar and pipe bomb	0	Check for tightness/Paul
Climb onto scaffold and insert bomb in pipe end. Fit bomb restrainer/catcher. Watch out for drips	10	Wear safety harness. Wear acid-proof gloves, wets, monos. HAZIDs and site regs./ Peter
Fit splash shield if practicable to deflect discharge into bunded area.	10	Get help to lift shields. Secure shields adequately to contain jet./ John Paul
Attach hose to Rotofan and stretch out hose, connect air line, check oiler. Back off speed regulator	20	Monitor each other, fit coupling clips and restrainers. Prestart checklist/ Paul
Inform client and other trades of your intention to start up	0	Prestart visual site inspection. Walk around checklist/Peter
Start pump at half pressure and check for leaks. Depressurise to attend to leaks	18	Do not handle hose while checking, visual check only/Paul
Start up and run to full pressure, start rotofan	12	Stabilise Rotofan—prevent roll over, monitor hose for leaks/ Paul
Check spray out of pipe, where is it going?	12	Watch for windblown waste, other trades and equipment, monitor bunding and vac-truck activities/Peter
AND SO ON—YOU CONTINUE FROM HERE		

Figure 2.4. Typical safety work plan

Hazard Identification and Treatment

Checked by: _____ Dated: _____

	CHECK STEPS	Yes	No
1	Is everyone inducted on site and properly trained?		
2	Are all staff fit and alert?		
3	Do all staff know their function for the job?		
4	Do all have the correct PPE and wearing it?		
5	Are the pump and engine prestart checks done?		
6	Have all hoses and components been inspected, fit?		
7	Are all hoses secured and restrained as required?		
8	Is the barricade tape up and secured?		
9	Have you checked floors above and below you?		
10	Have you fitted all signs?		
11	Have you attached your Personal Danger Tags?		
12	Have you checked all isolations with client?		
13	Is a site walk around to check for hazards done?		
14	Have you managed those hazards?		
15	Have you tested emergency phone and procedures?		
16	Have you checked the function of eyewash/showers?		
17	Is the site clear of obstacles (housekeeping)?		
18	Have you investigated spoil runoff/disposal?		
19	Have you checked the exhaust discharge location?		
20	Have you advised other trades of your intentions?		
21	Have you checked water runoff result under truck?		
22	Have you checked wind direction and splash travel?		
23	Have you informed the client of water usage?		
24	Have you got the client's permission to start?		
25	Have you checked for leaks and fixed them?		
26	Are all hose ends restrained?		
27	Have you established a signal procedure?		
28	Have you checked nozzle configuration?		
29	Figure 2.5. Prestart site checklist		

Figure 2.5. Prestart site checklist

By breaking down the job into simple steps, it becomes a simple job to monitor safe activities, forces you to rate the hazard and to reduce its impact, and provides a step-by-step activity list that prompts you to undertake certain tasks at certain times.

After the job, always review your HAZIDs and work plan, you may well use it again at another time. Add in anything that cropped up that you did not pick up, and make the changes needed. These documents are "living" documents in that they can be changed and improved on each time you do the job. You might care to highlight those items with numbers in the dangerous level to ensure you take specific notice of those.

If you have answered yes to all steps in the checklist, you can proceed to do your job without further delay. If you have answered no to any item, fix it now.

The object of a checklist is to jog your memory; it is not that your management thinks you are an idiot. A checklist reminds you to do something. In the heat of the moment—when there is noise, a complaining client pushing you to get on with it—you can forget something. Anybody can. So use a list and tick it off as you go.

On-site checklist questions:

1. Is everyone inducted? Do not try to "sneak" someone in because you are short a person; get him inducted. The axe will fall on your head.

2. Are all staff fit and well? This is important; if one member of your crew was up late last night and is hung-over or doped up, he is a hazard liability to you. Get him or her off the site, and replace him or her. You might get hurt getting that person

Hazard Identification and Treatment

out of trouble. If, as the supervisor or charge hand, you knew that person was unfit and you allowed him or her to work, you are legally liable in the event there is an injury. Does anyone suffer from asthma or is a diabetic? You owe them extra care; ensure that their affliction cannot cause an incident.

3. Does everyone know what they are going to do? This should have been sorted at the toolbox meeting, not now.

4. Physically check each other for the correct PPE, it is easy to climb out of the truck and forget a hard hat or glasses.

5. Carry out the checks. Do not sit in a warm comfortable cab and tick the boxes. Go physically and *check*.

6. When you walk your hoses, run them through your fingers. Does a wire spike you? That hose is not fit for use. Remove it from the line and tag it out for repair. (One broken braid requires the hose to be taken from service.)

7. Ensure that safety stockings are correctly secured and hoses cannot rub against steel work, egg grating, or hot pipes.

8. Barrier tape should be about 1 metre high at the centre sag point and no more than 1.2 metres high at each end. It must take a take a conscious effort to pass under or over your tape. A person should not be able to step over your tape. Signs indicating high pressure water jetting work in progress should be placed in normal access ways.

9. Check the floors above you for work above you that may be a hazard to you or your workers: tools falling through, weld or cutting torch spatter, falling materials, and so on. Check below you for signs that persons are working there; you don't want to dump a layer of acidic waste product on top of them. If in doubt, barricade and sign out these areas.

10. Site walk-around. Take a walk around, outside your immediate work area, and check to see if there are any things or activities taking place that may become a hazard. Once working, you are concentrating on the job and do not have the ability to watch 360 degrees. During shutdown situations, there are often cranes lifting loads above you. Talk to the crane operator and explain that you will move, but that he must advise you of a potential risk or hazard in advance.

11. Test the emergency phone, and discuss procedures with each other. Find the nearest emergency phone; tell your crew where you are. In the event you need help, "near the blue tank" does not help the ambulance driver. Locations such as "at the junction of A Road and B Drive" or "area 45 tank 16" do.

12. Test the function of eye wash stations. Always test run the eyewash and emergency showers before starting work in a chemical or acid plant. Make sure the water is clean and cool. On some sites, the emergency shower and eye wash is hooked up to an alarm system. Pick up the emergency phone, and tell the operator you are going to test run the shower. Or you could well end up red-faced when the ambulance and first aiders arrive.

13. Housekeeping is essential. In the event an operator needs to get to the eyewash station, he will be running with eyes closed. If there is a pile of scaffolding tube in the way, he may fall over it. Keep access and egress ways quite clear.

14. Remember *you* are responsible for your pollution. If your spoil is running to drains, check that this is legitimate and if possible get a work order to state this *and* get it signed by the client.

15. It is easy to forget the pump engine's exhaust blowing out of the top of the pump container. A recent incident caused by exhaust gas blowing onto a cable tray cost a contractor

Hazard Identification and Treatment

more than 1 million dollars. Blowing exhaust gas into a ventilator for an office block is not very friendly. Think.

16. Noise and spray is pollution; you are responsible for it. Ask other trades to wear glasses and hearing protection.

17. All water jet pumps seem to end up with water on the floor underneath them. Sometimes this can create a hazard—address it.

18. Be aware of windblown spoil. The water coming back out of the pipe is pretty well vaporised, but it still has the ability to carry contaminants that can be carried by the wind onto cars, buildings, or even people. The author once had to detail thirty-one cars in a car park five hundred metres away after covering them with a fine film of caustic.

19. Are you are hooked up to the fire hydrant? You are about to draw a heap of water. Most hydrants are alarmed, and when a significant amount of water is drawn, things begin to shut down. Tell the client how much you intend to draw.

20. Tell the client and other trades in the area that you are going to start. If others are working nearby, issue earplugs or advise them to fit muffs. Noise is pollution, and you are responsible for your pollution. Don't forget you have a duty of care for others around you. If they refuse to wear suitable PPE, you may consider discussing the problem with their supervisor.

21. Establish a communication procedure between the hose or gun operator and the observer. With noise and sometimes distance, it is not always possible to shout at each other. Establish a signal system that does not require voice to be transmitted.

22. Do you have you the correct nozzle for the job fitted? This is an essential activity. If the nozzle was on the end of the hose

when you took the truck out, there is a good chance that it is not the nozzle you need. We discuss configuration later in this book.

23. Most good operators know what is being done, and the observer *is* observing. If something goes wrong, the system can be rapidly shut down. It is essential that you establish a basic signal for things you might need: shut down, more pressure, less pressure, and so on. It is illegal in Australia for a person holding, directing, or in control of a nozzle not to have the means to shut it off. This may be a trigger or a pedal.

24. Once again, check your tooling. Make sure you have the right tool for the job. You are there to do a job as efficiently and professionally as you can; do it right first time.

Chapter Three

Being a professional

Let us begin at a logical point, with the reason we are employed: the customer. Without the customer, you and I have no job, no income, and no security. Let's have a look at this faceless nuisance and pain in the neck.

The first thing to remember is that he hasn't got a clue about high pressure water. He has no concept of jet velocities, reactive forces, and nozzle sizes or hose capacities. In most cases, he is a little suspicious of our equipment and us. He doesn't like to pay the huge amounts of money we charge and would love to catch us out. Not knowing exactly what we are about leaves him uncomfortable and uneasy. It takes very little to upset him under these conditions.

He is exactly the same as you and me. He has a family and a job he grumbles about. He gets paid less than he thinks he deserves. His wife doesn't understand him. His boss gives him a hard time, and he has his off days.

We are all customers in one way or another. We all expect to get value for our money. When you buy a carton of milk, you reject the damaged or dirty pack. If the movie is bad, you grumble and say, "I ought to demand my money back". If your children's shoes wear out in a few weeks, you will feel cheated by the manufacturer. If the garbo spills rubbish on the side of the road, you complain about paying the rates. Join the club, so does your customer. He is human too.

By requesting our services, he is offering to pay money to have a particular job done. Like you, he expects the job to be done efficiently and neatly and to have the site left clean and ship-shape when you leave.

His requirements are not unreasonable. If they were, we wouldn't work for him, and he would not last long in his job.

When he allows you onto his work site, he is extending a compliment to you. He thinks you are good enough at your job for him to feel comfortable with the outcomes of your actions.

The word *comfortable* is worth taking into your vocabulary. Then take it out and look at it now and again. Am I comfortable with the quality of the job I am doing for my customer? Will he be comfortable with the finished product? Think of him driving home thinking, *The boys did a good job again.* He will be comfortable thinking about you. Uncomfortable could sound something like this: "I wish I'd never let those morons near my plant. What the hell am I going to tell my boss at the morning meeting?" That's UN-comfortable.

The client is uncomfortable, and you are uncomfortable. You look at your wife and children at the supper table and think, *Your tired old dad screwed up today.* Now, there is real discomfort. When you look in the mirror in the bathroom, are you comfortable with the fellow you see there?

It's payday. You stand in line and take your pay. Are you comfortable?

I am not implying that you are not good at your job. You are most likely very good at it. We all have off days; let's just keep them to a minimum and consider the customer. He *is* a human being, doing a job just like us.

Let's have a look at what the customer expects of us:

Value for money. He is paying by the hour, in most cases. Give him sixty minutes in that hour and do not "cook the books". A job done on time or in less time will practically guarantee a repeat performance. Repeat performances keep us employed.

Give it your best shot. Do not get casual about the job. Try to improve your performance every time you repeat the job. Make it a personal challenge to improve; try a new gadget or change the angle. Go on, be adventurous; Columbus was. Too many of us stand in line to do our time, hanging onto the gun or hose, but how many of us consider it fun? It can be. Be challenged; take a chance; try something new. The only person never to make a cock-up is the person who will not try. If it works, you will be a bit of a hero. If it doesn't—well, you tried; no one will fire you for that.

Good housekeeping is essential. Clean up after the job. Do not leave the site as you found it; rather, leave it as you would like to have found it. This is particularly so when related to safety. You may have found the manhole open when you arrived, but if you leave it open when you leave, and some idiot falls in, whose fault is it? Yours. If you use the site washrooms and toilets, behave as if you were at home; rinse out the hand basin, and put the paper towels in the bin. Aim to please. People notice. Why do you think so many sites have toilets for contractors, and contractors cannot use normal site facilities? We are condemned for the actions of a few.

Observe site safety rules. Wear the protective clothing the client requires you to wear in the manner he requires it worn. Barrier tapes, warning signs, and notices are put up for the attention of *all* persons working on the site. As a contractor, you are not immune.

Keep the customer informed. If he knows what is going on, he can feel comfortable about the job. If his boss asks him what is going on and he does not know, he looks bad. You do not have to tell him you might have screwed up, but you should tell him why you are delayed. You should not have to lie, but do give him a plausible and understandable reason for a delay or breakdown. Someone might ask him, and he will feel very uncomfortable if he cannot answer.

Never belittle your own company, management, or service department to the client. It is no joke to him, even if it might seem amusing to you. He is paying for the "joke" services.

Never discuss prices with your client unless it is your specific job to do so. If he asks what the machine costs per hour, say, "It depends on the job", and refer him back to the office. Prices do vary, and hourly rates are often varied between customers, depending on a number of factors. You could upset a carefully worked-out system.

Always know exactly your machine or truck weight, height, and pump capacities. Litres per minute, gallons per minute, pressure in bar and psi, engine horsepower, fuel consumption rate, noise levels, and so on are important factors to the client. He needs such data to calculate extra draw from his fire mains, drain capacity, treatment works, ground loading, site safety, support, and so on. Do not go on site unless you know what you have on the back of the truck. You may look an idiot, and that's uncomfortable.

Do not discuss company policy or business with the customer or his employees. A sure-fire way for any skeletons in the cupboard to get back to the customer is to tell it to one of his workers in confidence.

Take particular care that you do not cause a demarcation dispute between one union and another. If you have to cross the fine line, do it discreetly and without any unnecessary fanfare. If you have to wait for a rigger to take a pulley down, do so, log it in your daybook, and try to get it signed off as waiting time. Do not take the thing down yourself. The potential dispute could have you and your company banned from the work site.

Obey the client's site road rules. Your truck or van has your company name all over it. They know who you are, and someone will see the "crime". It will surface one day when you least expect or need it. Contravention of a road traffic regulation indicates a general disregard for safety rules. Being placed under the microscope by the safety officer may cause you considerable discomfort.

Do not "borrow" anything. You may think you are doing your employer a favour by begging or trading a drum of diesel for some stickers. But being in someone's debt can cause serious drama later and can also make you an accessory to a crime.

Think before you act, remember the "comfort factor", and always treat your customer as you would like to be treated. After all *he pays your wages.*

OTHER CUSTOMERS AND THE PUBLIC
The people at your depot are also customers in one way or another. The fact that you do not pay for a service makes it no less deserving of consideration. The fitter who fixes your pump is providing you with a service, and you are his customer. If you are not happy, you are an uncomfortable disgruntled customer. After all, in the long run, you pay his wages through your production in the field. If he hands the truck back with grease all over the steering wheel you, his customer, are not pleased. The person who provides any service is a supplier and a customer too. If you don't pay for his services, try saying thank-you for a job well done.

"Thank you" is a phrase that gives comfort to anyone who receives it. It also makes the giver feel good. Try making a point of thanking the workshops for maintaining your pump. The service you get next time will likely be out of sight.

Always check with your customer when leaving the site, and say thank-you for his help (even if he did not give any, he will feel good). This is common courtesy and good working practice. If you are working on a person's site under the instruction of someone else, always consider all the parties, and communicate with all the parties.

Smile, damn it, even if your boots are full of water and your back aches. A smile gives relief to the receiver and generates a secure, comfortable feeling. A frown indicates trouble and discomfort and generates suspicion and distrust. If you are tired, remember that a smile uses half the muscles a frown does.

DOWNTIME

Downtime is time spent not working because *you* are broken down or not prepared. Let's consider what downtime is made up of from your employer's and client's point of view.

Your time *costs* your company about 40 dollars per hour. For the sake of simplicity, we will say that your sidekick makes the same, so that is 80 dollars per hour in wages. The client has ten men waiting for you to finish the job. They are sitting, drinking coffee; that's 400 dollars per hour it's costing him. The job is holding up the plant production. The cost of this plant in down time is 30,000 dollars per hour. The cost of not being able to supply his customer with finished product multiplies out the equation, so that your 80 is in fact 80,000 dollars per *hour*. Is it any wonder the client is so uncomfortable he is pulling out his hair?

Downtime is 1 per cent accident and 99 per cent poor planning on *your* part. You can line up the excuses, but when you think them all through, you cocked it up—no one else did.

Being a professional

You knew full well that the pump had a seal going or the regulator wasn't working properly or the hoses were shot. You knew you had only half a tank of fuel or that the fan belts were cracked. If you didn't, why didn't you? It's your unit; you took it out of the yard. Why didn't you check it before you left? Did you ask what the job was? Have you got all the tools you need? Why not?

Here's how downtime relates to your company. When a company purchases an item of plant, they must realistically estimate the number of days or hours it will work per year. The result of this estimate is normally best case and worst case with an optimum. In most cases, plant owners estimate about half the year—that is, 20 days per month for 6 months, or 120 days per year. They add in a factor for wear and tear, your wages, time clerks, payroll taxes, sales, workshop labour for maintenance, insurance, leasing fees, and a management fee. Divide all of this by 120 to give a day rate cost. A profit of about 20 per cent is then added onto this figure.

For the sake of this example, we will use a day rate for the unit is of 2,000 dollars per day. This is 1,600 dollars cost plus 400 dollars profit.

You drive out to site, and the unit breaks down. Your wages still need to be paid, and the cost of the fuel, tyres, and wear and tear on the truck simply to get there and back adds up. The workshop has to change its planned maintenance schedule, and the client is unhappy. You not only lost the day's income, you are at least 1,600 dollars in arrears on overhead. If you make as much as 400-dollar profit per day, you must work for three days virtually for nothing to get the required *costs only* back. You are also down 1,600 dollars on your estimated/budgeted profits (four days at 400 dollars).

The profit margin provides a number of important factors, including bonuses, staff facilities, and company investment, with its resultant job security for its employees and future growth. Another point to remember is that another job is waiting; you've got to do that one too. You never catch up.

WAITING TIME

When you arrive on site, do not dither about. Get on with it; set up your equipment; lay out your hoses, barriers, and signs; get everything shipshape and ready to roll. Do it right the first time; make the site secure and safe before you start to work. If there is a delay on the client's part, you can begin to charge waiting time. Get the book signed by the client or at least initialled by someone in authority. This is particularly important if your client is working shift and someone else may sign you off.

It is very important to define clearly the cause of waiting time. Put as much detail as you possibly can into the book. Use names, if possible. There is always an argument about it afterward, and you are normally the scapegoat.

If any stoppage occurs that is not under your control, you may charge this to waiting time only if you are set up and ready to go. The client is required to pay you an agreed hourly rate. This rate should be worked out before the job starts, so the client is aware of it.

Do you know what the term *professional* means? Not much by definition, really. The definition in the *Concise Oxford Dictionary* is "of, belonging to or connected with a profession". In the real world, it is supposed to mean that the person being discussed is good enough at his job to get paid for doing it. In fact, it implies much more than that. It implies dedication, concern, striving to do better, an educated approach, and a job well done that you would be proud to sign your name to.

So let's consider our professionalism as viewed by the client.

The truck drives on site. Are you late again? Is the truck clean? Is everything in good order? Is there any oil leaking out the bottom? Are there untidy bits hanging off? Are there clouds of smoke from the exhaust? Is the cab filled with food wrappers and empty drink cans? Any of these, and we have already gotten the client thinking uncomfortable and "unprofessional" thoughts.

Do you climb out of the truck wearing clean overalls, or jeans and a T-shirt saying something inane? Your client is not the least bit interested in which brewery you support or whose motorbike you ride. Are your overall sleeves still attached and your boots clean and in good repair? Is your hard hat on your head *minus* stickers and stupid comments? Are you safety glasses or goggles on your face not in your pocket, cubby, toolbox, or kit bag?

Do you know you'r client's name? Why not? Stick out your hand and introduce yourself. Tell him who you are and where you are from. It may be all over your overalls, but no matter. Tell him who at your office gave you your instructions. Confirm that the instructions you have received match the job he wants done. Ask him about special site conditions, gas, power, hazards, poison substance, and so on. Confirm the disposal of waste (anti-pollution laws make you responsible for your runoff, not the client). Make him feel that you know your job, are a professional, and are interested in his problem. In fact, make him quite comfortable having you to solve his problem.

Ask him if there is any information he would like. Does he require periodic job status reports? More often than not, he will not, but it makes him feel good to say so. Comfort, comfort, comfort.

Do the job properly. Take no shortcuts. Obey the safety rules. Clean up. Close hatches and manholes and *remove any personal safety tags*. Remove your barrier tapes. Load up the truck, and drive a few yards away. Stop, get out, and look back to make sure nothing was left behind. Tell the client the job is complete. Get the book signed, and leave the site.

If you are not *100 per cent happy with your finished job*, though you have given it your best shot, call the client and explain—*before* you pack up—and let him approve the job. If he wants more and you are sure you cannot give more, do *not* argue. Call your superior (he's on duty twenty-four hours too), and get him to sort it out.

(A tip on tagging: When you tag out a particular piece of machinery, tie a blank tag to your truck steering wheel. It is difficult to drive a truck with a bunch of tags tied to the wheel. Six blanks? Go find six tags. Forget, and you are in serious trouble.)

If you are working in a shutdown window and have a pipe to clean, for instance, seriously consider the possibility of getting it completely clean in the time frame. If you are in doubt, clear a hole right through from end to end and then work at enlarging it. If you get pulled off before you have completed the job, you have given the client back 50 or 60 per cent, or whatever, of his flow. If you had cleared half the pipe 100 per cent clean, you will have given the client nothing. Your client is most likely engineering or maintenance. His client is process. Process rules all plants, and it is smart to think a little bit ahead.

Talk to your client if you feel you may have trouble completing a project. Make him the alternative offer rather than turning down the job. Always think and act positively. There is no such thing as *can't*; there is only *we'll try*.

As a recap, a professional approach requires you to consider the following:

BEFORE LEAVING YOUR YARD—DO THIS
- Check your equipment (both truck and pump), including the oil, water, fuel, tyres, indicators, lights, mirrors, hoses, fittings, supply hoses, filters, nozzles, etc., etc., etc.

Being a professional

- Understand clearly what it is you have to do. Find out. *Never assume.*

- Distance from water supply to truck

    - Distance from truck to job site

    - Is the job on ground, off scaffold, or what?

    - What PPE do we need as standard?

    - Do you need any of these:
        - Safety harness
        - Breathing apparatus
        - Waders
        - Special protective clothing

- Read the Job Safety Analysis sheets.

- Read the Hazards Identification Sheet.

- Discuss the job with your crew; anticipate the worst and prepare for that.

- Anticipate the client's misinformation. He will always exaggerate a problem up or down; it will always be more or less than he claims. Allow for this by carrying extra tools.

- Make sure you know the client's name, contact address/phone, site location, and title.

- Check your appearance and that of your equipment. Take a minute and tidy up.

ON ARRIVAL AT THE JOB SITE—DO

- Wear the prescribed safety/protective clothing. If your company requirements are greater than the client's, obey your company. If the client's are greater than your company's, go with the client.

- Report directly to the client. The client is the person who contacted operations and ordered the job.

- Confirm your instructions, job procedure, and site conditions with the client. Any variation that may involve you being on site longer than anticipated must be communicated to your operations controller or client service representative at the earliest, preferably prior to starting work. Do not argue with the client; let someone else do that. You may politely point out that your instructions differ from his expectations. That's it.

- Inform your client of your water consumption, axle weight, and noise levels. A truck-mounted, engine-driven pump unit is +100 decibels to 10 metres; a nozzle up to +120 decibels to 6 metres. Inform him you will barricade off the work site and that eye/ear protection must be worn inside the barrier or within your recommended distance. You know your job; you should know how far your spoil will go and how loud the noise will be. Eye/ear protection should be worn by anyone working within 40 metres of the nozzle outlet.

- Set up on site, and fit barriers and signs (all job sites). Protect all normal access ways. Barriers should be at least 6 metres from the nozzle outlet. They should be set up so that that they are 1200 mm high at each securing point and no less than 1,000 mm high at centre "sag" point. Each site has different tape requirements. The site safety officer should be asked what style of tape or barrier he requires.

- Keep hose lengths as short as possible and protect them from damage because of
    * vibration
    * hot pipes or machinery
    * traffic

- Keep the client informed.

- Log all client-related hold-ups, detail the time, and the reason(s), and get a name to refer to.

- Once you are set up on a job site and waiting to start but are prevented from starting—for whatever reason—the client pays.

- If you are not set up, he may claim he was waiting for you.

BEFORE LEAVING SITE—DO

- If in doubt as to the quality of your job, inform the client of your completion before removing your equipment.

- Get the client to view the job and approve it.

- Clean up, close all manholes and trap doors, and remove debris to client-prescribed places.

- Remove barriers and signs. Place used barrier tape onto a reel for reuse.

- Drive a few yards away, stop, get out of your truck, and check that you have left the site safe, clean, and tidy. Is that a water jet lance leaning against that column?

- Inform the client you are leaving the site.

- Inform company operations you are leaving the site.

- Fill in the logbook.

Chapter Four

Understanding and Checking Your Equipment

How many of us know the pressure and flow capacity of our pump? When I ask operators in a classroom situation what the performance of their pump is, less than 10 per cent know. That, to me, is scary. You have a tool that can seriously injure or kill you, and you do not know what its performance data is.

It follows that if you do not know that, then you most certainly do not know what the Safety Pressure Relief Valve is set at. Without that, how do you know whether the hose, gun, nozzle, or fitting you are using is suitable for the pressures your pump puts out? Very few operators know the maximum working pressure of the hose that they wrap around their body or pass between their legs.

I often get the comment "That is the boss's/operation's/workshop's problem". It most certainly is not. You are using the gear; you are responsible for it. Get to know it extremely well. It's what makes your living and stops you getting hurt.

High Pressure Water Jetting – An Operator's Manual

Write *in big letters* inside your truck:

Engine power: xxx HP (divide HP by 1.4 to get kilowatt)

Pump model: xxx

Pump maximum capacity: xxx litre per minute at xxx bar or psi

Relief valve/bursting disc set at working pressure of xxx bar (should burst/relieve at no more than 20 per cent over the Maximum Safe Working Pressure [MSWP] of the weakest component in your system)

Unit total weight on the ground: xxx kg/tonnes

Unit total height and width: xxx mm by xxx mm (minimum clearance)

You might like to write your fuel consumption per hour. Ring up the engine manufacturer and tell him what engine you have, and he will tell you how much diesel fuel you should consume in an hour. Now you know when you need to fill up.

He might ask you what horsepower you are drawing. This is easy to calculate: bar x litres x 0.002 : 0.746 = horsepower of your water. Multiply that figure by about 1.5 to get the power needed to run your pump.

Example: 500 bar x 90 litres x 0.002 : 0.746 x 1.5 = 181 horsepower

The fellow may tell you that your engine consumes 0.3 kg of diesel per horsepower per hour. So, that's 181 x 0.3 = 54.3 kg per hour. As near as matters, that is 50 litres per hour. Got a 300-litre tank? That's about 6 hours running. Under normal working conditions, you are unlikely to run your engine for much more than 6 hours per day, so that is one day's fuel.

Understanding and Checking Your Equipment

You might add to the list on the wall of your truck the type of oil in your engine (it's not smart to mix brands). While you are at it, list the oil your pump uses. Got a hydraulic power pack on board? List its oil too. It is not difficult to do the things expected of a professional, is it?

As a responsible operator, you need to ensure that your equipment is in good order, so check that the oil is topped up. Your truck is constantly using engine oil. Expect to use up to 1/2 litre of engine oil for every 100 litres of diesel fuel—or, as in the case above, about 1.5 litres per tankful per day). If you are not using oil, you are diluting your oil with fuel. Report it as soon as possible; it needs to be attended to. Oil diluted with diesel does *not* lubricate your engine.

The Pump

In most cases the pump we use is what is known as a triplex pump. It has three plungers attached to a crankshaft, which strokes each plunger through one revolution. Typically, a pump rotates at about 400 to 500 revolutions per minute. A typical engine rotates between 1800 and 2200 rpm. It follows that somewhere in between is a reduction device reducing the speed by four times, or 4:1. In some pumps, this is done via a big pulley and a little pulley with as many as 6 V belts on it (write up the V belt part number). In some cases, there is a gearbox mounted between the two (which takes another type of oil, so make a note.). In others, the reduction is inside the pump with a reduction gear driving the pump's crankshaft.

The pump is fitted with a Safety Relief device, either spring loaded or bursting disc. The bursting disc, as a Safety Device, is gradually replacing the spring-loaded valve. All styles of Safety Relief Valves are clearly marked with their relief pressure; you MUST know what that pressure is. When you buy replacement bursting discs each comes with a metallic tag or sticker which details either the burst or maximum safe working pressure (MSWP) for that disc. Stick it onto the body of the holder for reference, that's what it is for.

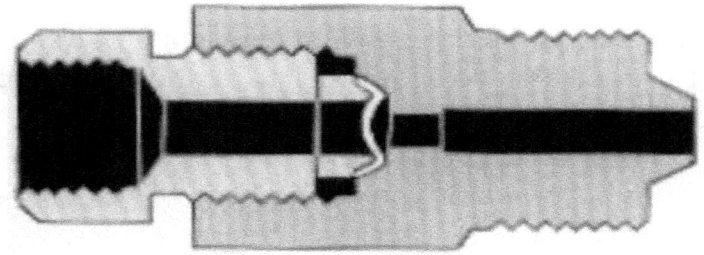

Figure 3.1. Burst disc holder

The spring-loaded safety relief valve relieves at the set pressure (should be set and sealed at 20 per cent above the maximum working pressure of the *weakest* component in the system), but some makes hold the pressure at the set pressure, working a bit like a regulator—that is, if you have a problem, the pressure does not fall off, but is maintained at relieving pressure. If you are at the end of the hose in trouble, full pressure can still be coming out at the "bite you" end. With a bursting disc, the disc bursts and dumps all the pressure in the system until you turn the pump off.

If a safety device activates, it is as a result of something *you* have done wrong. You are not permitted to fiddle with safety relief devices; they are not operator adjustable. The relief pressure is engraved onto the body of the valve. Read it.

Most pumps today are fitted with some form of pressure or flow regulator. These *are* operator adjustable devices and are used to control the pressure going to the nozzle. They include the following:

- Computer set and monitored via keyboard and screen

- Hydraulic, and set by an oil pressure regulator

Understanding and Checking Your Equipment

- Pneumatic, and set by an air-pressure control valve

- Adjustable spring tension, manually controlled

When setting these devices, always ensure that the gun or nozzle operator is in total control. With the trigger pulled when using a dump gun, gradually close off the valve until the nozzle operator indicates enough pressure. Always keep an eye on the pressure gauge to ensure that you do not overpressure the system.

When using a pressure control regulator, always start the pump with the regulator backed fully off or open. The pressure regulator depends on a spring to load up a valve that opens at the set pressure and dumps the water to the ground or back to tank. As the restriction (nozzle or gun trigger) increases, the valve compresses the spring and allows increasing amounts of water to dump. This device is used to maintain set pressure at the nozzle. Be aware that water passing back to the tank via the regulator is wasted fuel; that water should be coming out of your nozzle.

As the nozzle wears, the valve gradually closes, maintaining a constant or set pressure at the nozzle. These devices are designed to dump the entire pump flow if you let the trigger or the foot pedal go, the pressure or water at the nozzle stops instantly. The regulator opens right up, compressing the spring, and maintains set pressure in the system until you pull the trigger again. These systems are known as dry shutoff.

The flow control regulator works not unlike a tap; you start the pump with the valve fully open and close it off (trigger pulled or nozzles fitted), watching the gauge, until you reach the required pressure. Water can constantly dump through the valve; remember, this is wasted fuel. As the nozzle wears, the valve does not compensate, and the gauge pressure will drop off. This device cannot cope with a dry shutoff gun or pedal and is used on systems known as Wet Dump. When you let the trigger or foot pedal go, the water dumps out of the gun or foot pedal dump port, and pressure in the system is released through the trigger or foot valve activation.

NOTE: If you try to start a dry shutoff pump with the regulator fully closed or set at highest pressure point, you *will* overload the starter and most likely destroy it. At the same time the system is "live" when you are not in total control. This is often overlooked when you have had a system shutdown for some reason or another—typically low supply-water pressure. The pump shut down, you fixed the problem, and then you tried to restart without easing off the control device. So, when you stop work or before starting the engine back off the regulator, make this a habit. Pressure control regulators should not be left idle or parked up with the valve set tight; the spring will lose its tension and stop controlling. Be aware that dry shutoff units can maintain pressure in the hose line against a shutoff trigger or pedal for several days. *Always* dump the system after the engine is turned off.

The Pump Water Pressure Gauge

Pressure at the pump head is indicated by a pressure gauge mounted onto the pump head. The operative words here are "at the pump head". The gauge is not a crystal ball and cannot tell what the pressure is coming into your nozzle. We will discuss the pressure at the nozzle later in "Friction Losses".

Gauges should be selected so that the normal working pressure of the pump is read at about half of the gauge's capacity. In other words, a gauge fitted to a 1000 bar (or 15,000 psi) pump should be capable of reading almost 2000 bar (or 30,000 psi). Gauges are very expensive. If you don't load them up, they will last for years. If run at the extremes of their capacity, they will soon fatigue and fail.

Most gauges use a Bourdon tube to create the movement of the indicating needle. A Bourdon tube is, in essence, a copper or steel tube coiled like a flat spring and closed at one end. As water pressure is applied into the tube/spring, it tries to straighten out. This straightening motion causes the inner, or closed, end of the tube to rotate slightly. This rotation is transformed to a pressure-reading needle through a series of small gears attached to the end of the tube.

The pumps we use are normally triplex pumps (three plungers), and the pulsations caused by the three plungers develop a surge of pressure as each plunger strokes to its limit. (See "The Engine and Pump Unit" below.) Depending on the pump speed, this figure may be as much as 100 bar (1400 psi) per stroke. Between the actions of each plunger is a period of pressure reduction or variation (see "Slippage" below). This variation at every plunger stroke (about 1,200 times per minute) can cause the Bourdon tube to break from fatigue and destroys the small gears fitted to the indicator arm. To smooth out this pulsation, a pulsation damper is fitted between the gauge and the pump head.

Gauge Snubber and Pulsation Damper
There are various styles of snubber, all of which are designed to smooth out the pulsation caused by the plunger discharging into the three high pressure ports. Most are not expensive-looking items; in most cases, they resemble a hex socket. Have a peek inside next time you take the gauge off. The most common snubber has a very small hole, similar to a tiny nozzle, drilled through it; some have tiny sapphires inside. This small hole restricts the flow through to the gauge. Pulsations transmitted by the plunger are smoothed out by the action of the snubber, eliminating wide pressure variations into the Bourdon tube and gauge gearing.

Some snubbers use diaphragms with thick oil on the gauge side and complex flow-restricting or transmitting devices. These are very complex, and the pinhole one seems to last longest. The pinhole can become blocked with dirt and rust, and most have fine wire screens to filter the water entering the port. If you remove your gauge (and not the snubber fitting), do not start the pump for any reason, as the water passing through the wire strainer at speed will destroy it.

To add to the damping or smoothing effect, the gauge is filled with a clear, viscous liquid—usually glycerine. The glycerine is very thick, with a consistency close to 90-weight gear oil. This eliminates rapid movements inside the gauge housing and acts as a shock-absorbing fluid.

NOTE: Gauges that have leaked can be topped up with glycerine. This is a product your wife may use when icing a cake; it keeps the icing soft and shiny. You can buy it in most chemists or pharmacies.

Should the pump suck air intermittently, the pulsation can be greater than the snubber protection provided, and the gauge will fluctuate beyond the protection limits or capacity of the snubber. A damaged valve or seat will cause this fluctuation to occur. Rapid fluctuation will also damage the gauge. If you note the gauge fluctuating, you have a problem that must be attended to at once, or the gauge will be destroyed. A good gauge costs about a thousand dollars.

An indication of a damaged gauge is that it will not return fully to zero. This indicates that the Bourdon tube has fatigued and its "spring" tension has been lost or that one of the small gears has jumped a tooth. The gauge is no longer accurate, so you no longer know the pressure you are operating at. It is a standards requirement that a pump be fitted with a working and calibrated pressure gauge. There is no recommended periodic calibration made. But should the operator consider the gauge faulty, it should be removed for repair and calibration.

Understanding and Checking Your Equipment

Take care of your gauges; they are very expensive and are safety devices.

Pump Inlet and Outlet Valves

The style of valves fitted to a high pressure pump varies considerably. There are almost as many styles of valve and seat as there are pumps. The valves on a high pressure water pump are normally all pushed closed with a spring. The springs help the valve to close rapidly to enable the pump to stroke rapidly. A pulsing pressure gauge and a leaping hose often indicate a valve spring failure, thus indicating that the valve has not closed by the time the plunger is a good distance along its compression stroke. (A valve spring failure is commonly caused by cavitation; the bubbles have nibbled right through the spring. See "Cavitation" below.) The supply pump-fed inlet valve is normally a good deal bigger than the outlet that is supplied by pressurised water from the plunger or piston.

Once properly installed, valves can last for a long time. The only things that will destroy them prematurely are cavitation or dirty water. Unless it is time for an annual or periodic overhaul, replace only the damaged ones. *Do not disturb the good ones.*

To work out which valve is not working or closing properly, you need only to feel your hoses. If the high pressure hose has an uneven beat—light pulse, light pulse, *big* pulse, light pulse, and so on—this indicates a valve failure. But which valve, inlet, or outlet? Go to your suction hose and check to see if it is pulsing; it should not be. If it is, the inlet valve has failed. If it is not pulsing (only the HP hose is), it is a failed discharge valve.

To work out which cylinder or plunger the damaged valve is serving, take a long screwdriver, place your ear firmly over the handle, and touch the metal end of the screwdriver onto the head over each valve port. The noisy one is the damaged one.

The valves enclose the void space, preventing leakage of pressure back into the low-pressure or inlet side. The only place the displaced water can go is out of the discharge valve and into the pressure port in the pump head. In fact, there is another area where the displaced water may escape: the plunger seals.

Plunger Seals

There are many styles of plunger seals, and each acts slightly different from another. In most cases, the seal flexes and grips the shaft and the cylinder wall as pressure is applied. So, on the compression stroke, the seal grips the shaft with increasing force as the pressure builds. On the return stroke, as the pressure is reduced, the seal relaxes. It follows that the plunger shaft must be very, very smooth and clean. The housing where the plungers move has a cover on it. Keep it on.

Ninety-nine per cent of plungers depend on a small water flow to lubricate them. This water either leaks deliberately or is supplied via tubes from the supply water. Some have a small valve fitted to each seal housing so you can adjust the flow rate. Seals must drip a little water, not unlike a leaking garden tap. It should not be a constant flow; not only does that put pools of water in the truck, it also makes a mess underneath.

The seals enclosing the void space are normally fixed with the plunger running in and out through them. The seals fit tightly around the plunger to prevent water escaping out to atmosphere at the back. Seals wear out and have to be replaced periodically.

As the crank pulls the plunger back out of the void, the charge pump or hydrant pressure refills the compression chamber (void space) through the inlet valve. Each time the crank rotates, the volume of the plunger must be able to escape or something will burst.

Understanding and Checking Your Equipment

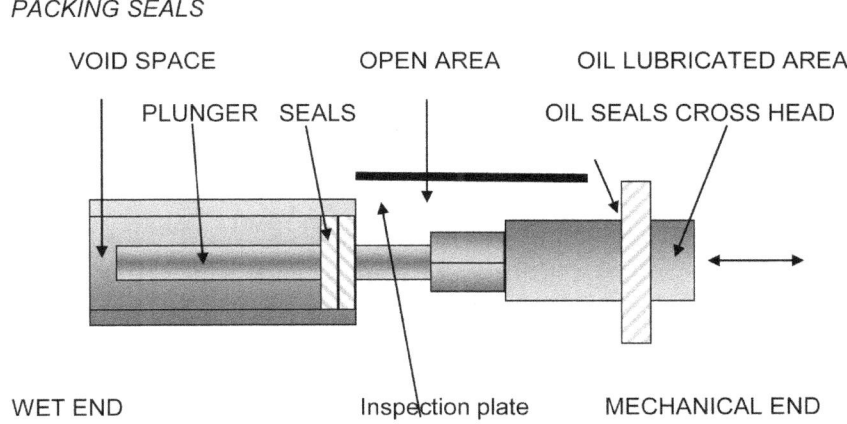

Figure 3.2. Plunger and seal concept.

Most pumps have an inspection plate mounted onto the casting which allows you to view the plungers through the open space between the wet end and the mechanical end while the pump is running. This allows you to check the plunger seals for leaks.

The water must *not* spurt out on each compression stroke. Water may spray across the gap between the head and the crankcase and get into the crankcase oil. Spurting water indicates a seal failure. Oil and water do not lubricate the bearings, and failure may be imminent.

Any colour change in the pump's oil indicates a problem, and the oil must be changed. Most horizontal plunger pumps have a rubber baffle disc mounted on the plunger shaft where the plunger joins the con-rod; these are important parts of the pump and must be there. When seals wear more and more, water leaks out. These discs cause the water to fall into the drain outlet, keeping it out of the crankcase.

Supply Water Filter

One of the major causes of pump damage and cavitation is a dirty or blocked water filter. The water filter is designed and provided to ensure that tiny particles of dirt are filtered from the water prior to delivery to the plunger.

Most pump installations have two inlet filters: a prefilter and a final filter. The most common high-flow prefilter is the Amiad Disc filter. It has a minimum filtered size of 120 microns. That can only just be seen with the naked eye. A blocked filter will not allow positive feed to the plunger. The plunger is forced to suck water through the filter element to get its needed fluid. Below we discuss cavitation in more detail, but for now, know that suction creates cavitation, and cavitation destroys pumps. In most cases, companies buy filters on price and not necessarily on performance. Check that your filter is capable of providing you with an unrestricted flow equal to your pump's maximum demand capacity with 50 per cent of its capacity blocked by dirt.

The bag filter used in the majority of pumps removes dirt down to 2 microns. Two microns cannot be seen with the naked eye. The Hammelmann labyrinth seal particularly needs very clean water. The bag must be inspected regularly; if any hole is visible to the naked eye, replace it. The correct way to check a bag is to hold it up to the sun or bright light and look in through the middle; if you can see a pinprick of light through the body of the bag, it must be scrapped.

Some pumps use a series of fine-mesh wire filters, each smaller than the last. This system is very good but expensive, and the feed to these pumps is normally very clean. Cleaning these filters is quite simple, a process of back washing is used to flush filtered water the wrong way through the filters.

Measure an unrestricted filter flow by placing a pressure gauge at each end of your filter set and noting the pressure

Understanding and Checking Your Equipment

drop through the screen or screens. A variation of 20 per cent indicates that it's time to clean or change the filters. A well-designed filter system has these gauges installed. Ninety-nine per cent of the pumps on the market today require a minimum 3 bar (45 psi) of positive water pressure delivered to the pump at the valves, at full working speed. Ideally you should have a water supply of 5 bar (most municipal mains and fire mains) at your hydrant or supply valve. Check most carefully that your hoses are not kinked and that no one can drive over them. Discourage people from walking on them, and most certainly KA anyone who stands on them for a giggle.

In some instances, pumps are fitted with small header tanks; in 90 per cent of cases, these are used to gravity-feed water to the plungers. Typically, there is not enough "head" height to feed the pump and valve correctly, and valve spring failure is a constant problem. If you are using a header tank, fit a small charge pump to your engine fan belts capable of taking water from the header and delivering it to the pump at a reasonable pressure. Ideally, fit two water feed systems. Where practicable, attach a hose from your supply directly to the filter inlet and allow water mains pressure to feed the pump. Where this is not practicable, switch over to tank feed. Most municipal fire mains are at 5.5 or more bar (80 psi), which is adequate if you use a 50 mm lay-flat hose, without kinks, for your supply hose.

Not only is cavitation caused by suction or inlet restrictions but, as the filter blocks up and suction conditions develop, dirt is also drawn through the filter plates or mesh by the suction, and other major damage is caused. Dirty water destroys seals and valve seats, and it destroys the water jet nozzle by abrasion. The use of ceramic and sapphire nozzles is very common in America and Europe. In Australia we dislike them because they fail so fast. They fail because we do not care for our filters and allow dirty water through the pumps. Clean filters give clean water freely to the pump. Keep them clean.

NOTE: Something to watch for with ultra-high pressure pumps' using sapphire nozzles is heavily mineralised water—that is, water containing dissolved minerals. These minerals can pass through a filter unhindered. Once pressurised in the 1500 bar plus range, the dissolved minerals revert back to their crystalline form, forming grains like fine sand. These are extremely abrasive and do a great deal of damage. If you have sapphire nozzles and are blowing out nozzles like crazy, though your filters are in good order, your problem is the minerals in the water. Use either demineralised water or reverse osmosis water. These grains rapidly wear out rotary or spinning nozzle holders too. Where a different water supply is not available, your management should consider charging the client a premium for the increased maintenance costs. This may force the client to look at his water and possibly supply you with water from the potable (drinking) water supply.

Slippage
If we calculate the flow of water through a pump by mathematical methods of plunger area multiplied by the number of plungers by the stroke distance and by the rpm, and so on, we would get a figure of 100 litres per minute, for example. This figure would most likely be accurate if we allowed the pump to operate at a low speed with no back pressure. This is normally the calculated figure used by the designer and shown on the pump brochures.

However, under pressure, certain things happen that the designers and manufacturers (who write the spec sheets.) ignore. In reality, when the pump operates at speed and under pressure with lots of horsepower pushing and with lots of friction trying to stop the water flow, we get a loss called slippage.

At the end of a pressure stroke, the discharge valve is open, and the plungers on either side of the one in question are returning. A short period exists in the revolution of the crankshaft at the point where the plunger begins to return when

Understanding and Checking Your Equipment

the discharge valve remains open. At this point, the pressure in the plunger void equals the pressure in the hose. Until you have an imbalance, the valve will not close, in spite of the valve spring. The pressure in the hose drops off as the water discharges from the nozzle, and at the same time the pressure in the void decreases as the plunger returns on its stroke. During this period, a certain amount of water returns from the pressure port into the void space. A similar phenomenon exists on the return stroke.

When the plunger returns, the pressure in the discharge line closes the discharge valve. The plunger draws back until the pressure in the void equals that in the supply or suction line (in most cases at 3 bar). At this point, the valve opens and water is driven in to refill the void. The plunger starts its return trip, and once the pressure in the void passes, the pressure in the delivery the suction valve closes. There is a short period when the plunger is moving to compression and the suction valve is still open. A small amount of water is returned into the suction port.

It follows that we have lost a fraction of our water from both ends of the stroke. This could be as much as 15 per cent. All of this assumes that the valves and valve springs are 100 per cent effective—maybe on a new pump, but not on anything that has worked over 150 or 200 hours. On the Hammelmann, the plunger labyrinth seals are wearing from the moment a new set is fitted. These too contribute to losses, which can be loosely called slippage.

Slippage is very likely pockets of air are hiding inside the head and associated pipe work. Air compresses (and expands again). It is important that air extraction devices are correctly located and that they work. Think about the location and whether all the air traps have been addressed. Get the air out before you pressure up; allow the supply water to flow right through the hose and run out clear before fitting a nozzle or gun and running the pressure up.

Cavitation

Any flow restriction can cause cavitation to a greater or lesser degree. Cavitation is the major cause of pump damage and costs thousands of dollars in maintenance and downtime.

> For every dollar we spend on maintenance, the pump needs to earn five dollars.

If your pump is running faster than the manufacturer designed it to run, it may not be able to get its required water flow in through the valves, cavitation will occur. The same will happen if the filter is blocked, if the hose is kinked, or if the supply pressure is inadequate. The pump is forced to suck to get its water supply. Vacuum then increases to a point where a portion of the water turns to vapour.

Water is a combination of two gases: hydrogen and oxygen (H_2O, two parts hydrogen combined with one part oxygen). It can occur in three forms, depending on temperature and pressure: liquid (water), solid (ice), or gas (water vapour). If we substantially reduce the atmospheric pressure by sucking, the liquid reverts to gas. Then minus 1 atmosphere—1 bar or 14.5 psi—is all we need to separate the gasses.

The gas bubbles, looking not unlike the bubbles on the side of a glass of beer, enter the suction chamber. When conditions turn from suction to compression, the bubbles implode (an explosion gets bigger and an implosion gets smaller, but the impact is the same) onto the metal parts inside the pump, blowing chunks of metal off the surface of plungers, valve seats, valve springs, and void spaces. This condition can often be heard through the pump casing; it sounds as if the pump is trying to pass gravel or nuts and bolts.

The resultant cavitation damage is obvious; it looks as if some creature has been nibbling on the sharp metal edges, giving them a finish similar to the cutting edge of a bread knife. Flat surfaces look as if something has taken matchhead-sized bites out of the steel.

Positive feed pressure on the supply side of all pistons or plunger pumps is essential to prevent cavitation. Plunger pumps must *not* suck. If you have any suction restrictions, you will cause cavitation that will destroy your pump. Suction hose selection is therefore very important, as is a clean filter. Any restriction to the pump feed or supply inlet will cause damaging cavitation.

 One of the most common causes of cavitation is insufficient supply water pressure. The best hose size for most pumps (to 250 litres per minute) is a 50 mm bore hose.

A simple check of the pressure in a soft rubber or lay-flat hose is that you should be able to walk over it without flattening it while the pump is running at speed. Your suction hose should lie still and not vibrate like the discharge hose does. Any movement on the suction hose indicates pump inlet valve damage; this should be attended to as soon as possible.

A common misunderstanding is that two 25 mm (1") hoses can be used in place of one 50 mm (2"). They cannot.

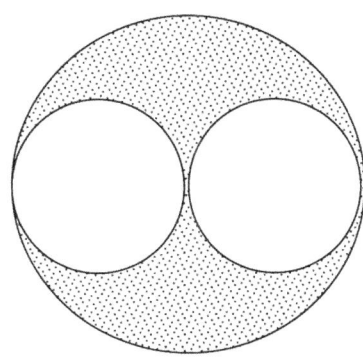

Figure 3.3. Hose sketch

The illustration above represents a 50 mm diameter hose. The two clear areas inside it represent two 25 mm hoses. The shaded area represents the volume not provided for by the two 25 mm hoses. Four and a bit 25 mm hoses are required to supply the equivalent to the one 50 mm hose. This calculation also applies to discharge hoses on the high pressure side: 2 x 6 mm does not equal 1 x 12 mm hose.

Area = πr^2

1" hose = 3.143 x 12.5 x 12.5 = 491 mm^2

2" hose = 3.143 x 25 x 25 = 1964 mm^2

For your information:

6 mm hose = 28.28 mm^2

12 mm hose = 113.15 mm^2

19 mm hose = 283.65 mm^2

25 mm hose = 491.09 mm^2

50 mm hose = 1964 mm^2

Another cause of cavitation is steps or reducers in your water line. This is not caused by the pump sucking but rather by water rushing over a step and creating a vacuum, not unlike a weedkiller sprayer you fit on the end of your garden hose.

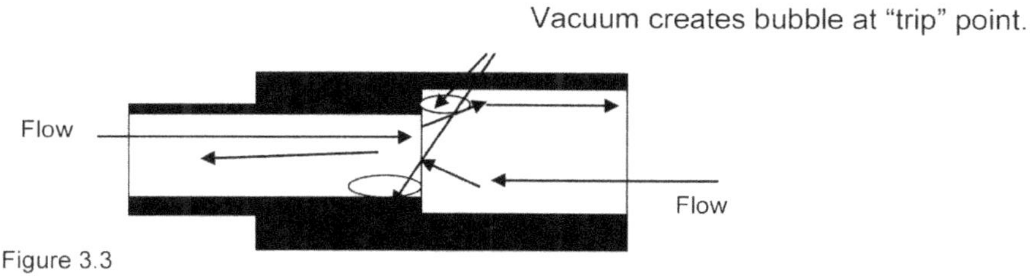

Understanding and Checking Your Equipment

Figure 3.3

Figure 3.4. Reducer fitting cavitation concept.

Water traveling at great speed through a ¾" fitting comes up against the flat edge of a ½" reducer. Water is forced upwards, and a tiny spot/bubble of vacuum is caused by the turbulence at the trip-up point. The bubble drifts along the fitting until the vacuum becomes pressure again, and a mini implosion occurs as the gas turns to water again.

This implosion bites a tiny bit of steel out of the wall of the ½" fitting. This is repeated, and the hole gets bigger until the fitting eventually fails, sometimes with catastrophic results. The same thing happens for water traveling in the opposite direction. Taper all steps in your pressure circuit, ideally 45 degrees. Take the time to inspect all fittings and joiners periodically; look inside along the bore for cavitation damage.

The Engine and Pump Unit
The engine that drives your pump works fairly hard when compared to engines on most installations. Most engine-driven installations have a constant and steady load, such as a generator or centrifugal pump. A triplex delivers three "peaks" every rotation.

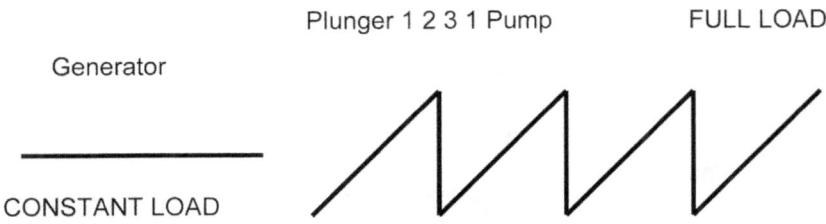

Figure 3.5. Load peak indication

The engine power transmitted through the coupling and reduction gears moves from almost no load to full load on average of 1,500 times per minute. At the same time, a four-cylinder engine is peaking with every firing stroke 8,000 times per minute. Imagine what is happening to the rotating components; they are taking a beating to say the least. This pulsation is transmitted to the coupling and key way, the engine and pump mounting bolts, the fan belts, the gearbox, and the crankshafts in both units and all moving parts.

To protect the rotating components, use the proper oil—a thin film of which is forced between the metal surfaces to prevent them touching and causing damage. It follows that your pump and engine need good oil, right? So, why is it that the engine oil looks like road tar porridge and your pump oil looks like coffee latté? *Any* change in the colour of your pump oil means you have a problem that must be fixed. The coffee-coloured emulsified oil (oil and water mix) means that you have water in the crankcase. This water has destroyed your oil, and metal might be touching metal, bringing you nearer and nearer to a major overhaul.

The engine oil does discolour; it turns from a nice honey colour to black within minutes of being put into your engine. This is caused by burnt fuel leaving carbon deposits (soot), which gather in the oil. The viscosity should not change. Study your dipstick when you have new oil in your engine, and see how it

Understanding and Checking Your Equipment

drips off. It should always drip the same. Smell the oil on the dipstick; it smells like oil. If the oil becomes runny, smell it again. If it smells of diesel, this is what is known as crankcase dilution; un-burnt diesel has run down past the piston rings and mixed into your oil. You are now lubricating your engine with diesel fuel—not oil. *Fix it.*

Your diesel engine must use oil. About 1/2 litre per hundred litres of fuel is consumed by a newish engine and as much as 3 litres by an old engine. If it does not, you *have* crankcase dilution and very unhappy oil.

What causes crankcase dilution? Not loading your engine, extended periods running at idle, or too big an engine for the job it is doing. Engine manufacturers design engines and fuel injection systems to operate continuously at about 85 per cent of their capacity. A 100 kw engine *must* use about 85 kw when it is working. The engine will burn only the fuel it needs; the rest runs down the sides of the cylinder, past the rings, and into the sump.

One of the biggest "costly savings" is hunting for a second-hand engine that nearly fits the power requirements. Normally one is found that is bigger than the required size, but it is cheap. Wrong. The damage it can cause plus the extra fuel consumption make it an expensive cheap engine.

If you cannot manage the pump's designed water pressure and flow, you are not doing your engine any good by running at half revs. Get a smaller pump combo or fit an extra nozzle. Make it work at the power and speed the manufacturer designed it to work at.

If you run it too slowly, it never gets hot. The rings do not expand, and oil can get up past them (grey smoke in your exhaust) or unburnt fuel can go down (and up, as black smoke in your exhaust). If you get a cloud of black smoke when you first start your cold engine, this indicates that the exhaust and manifolds are filled with unburnt fuel from that long idle period yesterday. If you get black smoke when the engine is hot (unburnt diesel), your injectors are shot and need to be overhauled. Grey smoke while the engine is running hot (oil burning) normally indicates worn or broken rings or worn valve guides. Either way, it looks unprofessional, and you are causing pollution. Get it fixed.

The pump oil does not get black and dirty but should remain clean, golden, and syrupy for most of its service period. If the oil becomes darker, it is overheating; look to your heat exchanger or improve ventilation. If it becomes the colour of coffee latté, it has water in it. This can come through from the plunger seals or from someone washing down the unit and squirting water in through the breather or the dipstick. No matter what the reason, the oil must be changed out now. The pump is a very expensive piece of a plant, and oil is cheap.

For every one dollar spent on repairs, the pump must earn five dollars to pay for those repairs.

Some pumps are splash lubricated, which means that the crankshaft in the pump's crankcase scoops up oil as it turns over and sprays it over the top of the crankshaft and the cross head slides. If the pump is running too slowly, it will not splash the oil over the cross heads, and they will rapidly wear and possibly seize up.

A recent calamity occurred when the battery had run flat so the operators left the close-coupled unit idling while they went for lunch. When they got back, the paint had melted black on the top of the pump crankcase and the cross heads had completely seized up—cost 32,000 dollar.

Others are pressure lubricated with a geared positive displacement oil pump; this too needs to run at a decent speed to keep the oil pressure up. *Think.*

The Coupling

The coupling joining the engine and pump needs to be monitored. If you are one of the unfortunates who have a fixed cover over the coupling, it is very difficult to inspect. The best way is to wipe the floor under the coupling with a clean rag wrapped around a long screwdriver. If the coupling is taking a beating, bits of rubber will be on the rag. Coupling covers are safety features required by law, so being able to get them off easily to inspect the coupling is an essential part of pump care and maintenance.

Typically, the coupling has a "pin and bush" style, and there should be no backlash—that is, the pins should not slop around in the rubber bushes. If they do, plan for a repair at the next service period. Couplings will last practically forever *if* they were put in properly and aligned. Any movement of the engine or pump mounting bolts will cause the coupling to be misaligned. With big engines, it is an investment to have the units professionally aligned. There are people who do this using laser equipment. Do not try to do this with a couple of straight edges. *It will cost you.*

Part of your daily checklist should include a visual inspection of the mounting bolts. It is a good idea to paint over the mounting bolts when they are properly aligned and fully tightened. Any movement will be easily recognizable, because the paint will crack.

The Clutch

Some pumps are fitted with an engine clutch, normally manufactured by Twin Disc. These clutches last for many years if properly looked after. Service includes regular greasing, but *do not over grease.* Typically there are four grease nipples on a manual clutch. Both of the lever arms can be greased fairly liberally once a week. For the thrust bearing (nipple located near to the inspection plate), use 2 squirts every 20 hours of operation, and for the pulley end bearing, use about 4 squirts every week.

If you over lubricate, the grease can spill over onto the clutch plate, causing it to slip and eventually fail.

The hand lever *must* have about 20 mm of play when the clutch is engaged. The movement at the top or handle end of the lever should be measured periodically. If there is no play, you are loading up the thrust bearing, and its hours are numbered. If there is too much play, the clutch will start to slip. The adjustment is fairly simple and requires a screwdriver only.

Take out the two screws holding the inspection plate, and remove it. If you are lucky, adjustment instructions are still written on the cover plate. Get a torch and look inside.

Look for a small, horizontal, spring-loaded pin that fits into a castellated grove in the outer circumference of the clutch plate. You might need to turn the clutch to get the pin to the top. Pull out, or push in, this pin until it stops and, using the screwdriver in one of the castellated grooves, screw in or out the plate by pushing the plate in a clockwise or anti-clockwise direction. Screwing the plate clockwise tightens up the clutch; anti-clock eases it off. In a worst-case situation, no more than eight or ten notches should be adjusted. If you need more than that, there is a problem that needs to be attended to by your workshop. Before you put the cover back on, give the moving parts in there a squirt of WD-40.

To ensure that the clutch works well for an extended period, *always* reduce the engine revs to idle before engaging or disengaging the pump. Watch the lever; it tells you what is happening inside. Having the lever too tight is similar to "riding" your car's clutch. If it is too loose, the clutch will not snap into place properly. You should not need to be Superman to engage a clutch; it should require not much more force than the weight of a couple of house bricks.

Diesel Fuel
Use clean fuel and keep it clean. How many times have we seen that statement written on engines, fuel drums, and sales brochures? How many of us try? Sixty per cent of field engine failures are related to fuel.

Handling fuel in the field is always a problem. We get it delivered from a contractor-site fuel tanker; we get it from a stationary tank; we get it delivered in drums; and we get it in jerry cans or 20-litre pails of dubious parentage.

Always assume the fuel is dirty; make every effort to clean it. The following are some fuel system tips:

If you are getting fuel delivered from a contractor-site tanker, discharge it via a filter; contractor fuel is notoriously filthy. If you don't have a filter, *get one*. Making up an emergency filter is a simple matter. If you cannot find anything better, fit a piece of grating about halfway down into a large funnel. Cover this grating with clean cotton rags which have been well shaken to remove any lint. The grating could be a piece of thin chicken wire folded over a few times and the wipe rag wrapped around it before placing it into the funnel.

The ideal would be to extend your filler pipe, fit a proper "high-flow" filter element into it, and fit the filler cap above that. Make it a fixed installation. This ensures that there is no entry into the tank except through the filter. If you can spring the funds, get a water trap too.

If you are getting your fuel in drums, always use a hand pump purchased with a filter attached. Never use a siphon hose; it is always full of sand and gravel. Buy spare filter elements, and change them out every ten or so drums.

Never store drums upright with the bung on top unless they are sealed with the supplier's seal cap (a tear-off tin cover over the bung). With the change in daily temperature, a certain amount of expansion and contraction happens inside the drum. This expansion escapes through a firmly screwed-down bung as the contents warm up. If it should rain, the "dish" on the top of the drum fills with water, and the contents cool and suck the water back through the seal into the drum.

If your seals leak when the drum is stored on its side, wedge a piece of wood under the upright drum so that the filler bung is at the highest point—above water level—should rain fill the dish. If the small bung has also been disturbed, tilt the drum even further with the bungs level (nine and three o'clock) across the top. Seal all empty drums, and store them lying down.

Do not put diesel into a drum other than one specifically reserved for diesel use. If you are not sure what was in it previously, or you are not sure if it is 100 per cent clean, do not use it.

If transferring fuel from any form of container, always run a little fuel through the hand pump and hose into a waste tin to make sure the hoses are clean. Take a look at the discharged fuel, and make sure it is not water contaminated. If it is, raise your suction 50 mm or so off the bottom and try again, making sure you clear the pump each time. It is good practice never to draw from the bottom of any container; dirt and water settle there. After passing a quantity of water through the filter in your hand pump, change it out. Some elements can be destroyed by water, and you could be pumping papier-mâché into your tank.

Another serious problem you may encounter is a bacterial infection in your fuel system. This is very common in site fuel

Understanding and Checking Your Equipment

tankers. The bacterium grows in water, under warm diesel, and looks like a black slime or fungus. It is a living organism and breeds rapidly. It coats fuel-line screens and filter elements, and it blocks up pipe work. If you get it, you will know it. Fuel filters last about twenty-four hours before the engine starts to starve and lose revs.

Before you allow the tanker to fill up your tank, run your finger around the inside of the fuel-hose handpiece nozzle; if it comes out with something that resembles black grease, that is the bug. So you have to take his fuel, because there is no option? Send someone to get fuel treatment additives ASAP, and dose the tank.

There are a number of bacteria killers on the market (such as Wynns), and it is good practice to add a little of this to your fuel every time you fill up.

NOTE: Do not store diesel fuel for extended periods.

The fuel manufacturers make a summer-weight and a winter-weight diesel fuel. The summer-weight fuel has added wax that in winter is likely to block your filters. Winter fuel has less wax and burns hot in summer. Diesel fuel is made to be used and not stored for longer than four weeks. This is serious, so stay aware of it, particularly if your unit travels around with drastic climate changes like coast to mountaintop.

Radiator and Thermostat
The radiator consists of a reservoir of water which passes through a series of thin copper tubes over which cool air is drawn by the fan. The cool air draws heat away from the copper pipe that in turn draws heat off the water. The hot water enters at the bottom of the radiator (from the engine) and returns back to the engine from the top via the water pump, which is normally belt driven and mounted on the same shaft as the fan. The water pump is normally an open-vein impeller centrifugal pump.

The radiator is normally so super-efficient at disposing of the heat that it could take the engine a great deal of time to warm up. We *want* the engine to run hot, because that keeps all the rings and pistons tight and helps the oil carry away deposits. To help it warm up quickly, the engine is fitted with a thermostat that closes off the water pump discharge and prevents the water circulating through the engine. The thermostat has a central hub like a copper bellows. As the water warms up, the bellows expand and open the thermostat door, allowing more and more water into the block.

This device is located where the inlet or top hose joins the engine block. There is a round, dome-like fitting, normally with two bolts holding it down. If you undo these bolts and lift off the dome, the bellows are visible. Take them out and wash them. Make up a new gasket, and replace the cap.

If the bellows and the copper tubes became coated with mud or slime, this could act as a form of insulation that could prevent the designed heat transfer, so the efficiency of the engine's systems would also change. Power would drop off, and fuel consumption would change. Oil temperatures could run high, and viscosities change. It follows then that the radiator must be kept clean and the water crystal clear.

Don't laugh. It can be done very easily if you have a clean system. Wash it out several times, fill it, run it, drain it, and keep going till the water comes out clear. *Do not use radiator cleaners* available from motor spares places; they do more harm than good and can destroy the cooling system. Bring the engine to heat point, shut it down, and drain it again. When the engine has cooled, refill the system with either a good grade of antifreeze or distilled water or rain water collected in plastic off a plastic roof (it should not have come in contact with metal or gutters, especially galvanized), plus a radiator protection additive. Tap water is filthy stuff; it contains all sorts of minerals and nasties that a radiator does not like. The system will remain sweet and will not corrode and fill the water with the usual brown sludge.

NOTE: If you are building a system, do not put the additive in until everything is 100 per cent okay. Additives are very expensive, and you do not need to leak all those dollars onto the floor. Make a plan to drain the radiator periodically by a tap you can get at and that you can fit a hose to for draining. You can reuse your additive/water mixture if it is still clean. Most radiator taps are stupid little plastic things placed behind a coupling or something that you can never reach.

Wash your radiator grid through with a garden hose periodically; direct the water in the opposite direction to the airflow. Some fans suck and others blow; check which one you have. Never use a water jet on the radiator; you will destroy the fins and possibly break a tube.

The Air Cleaner
Dry or paper filter element
This often-neglected piece of the engine plays a most important role in the operation, protection, and life of the engine. Its malfunction can cause excessive fuel consumption, overheating, power loss, excessive smoke in the exhaust, and carbon around the valves.

The air filter comes in two forms, either dry element or oil bath. Both do the job well *if* they are looked after.

Dry element air filters are more common nowadays and are made of a layer of porous paper with holes at about 5 micron. The paper is in folds to provide a large surface area. The circumference of the average big engine filter is about 1 metre. The length of the paper in that circumference when stretched out could be in excess of 20 metres. The manufacturer did this to provide an extended life and to ensure that there was *no* restriction on the airflow to the engine. Air flows freely through a new filter element. You could comfortably breathe through the air cleaner element *when it is new.*

The holes through the paper are smaller than visible pinpricks. So, it follows that it does not take much dust or dirt to fill up all the little holes. Because the paper has such a large surface area, 20 metres, it takes some time to block up. In an effort to further improve the life span of a filter, the manufacturers now provide two units—one inside the other. The first or outer unit has paper with holes to 25 micron and inner with holes to 5 micron, for example. In this way, the air is precleaned by the 25-micron unit (over 20 metres of paper) and then cleaned again over a slightly smaller length of paper. This may give as much as twice the life span of the elements.

There has been considerable discussion regarding the average size of dust molecules, and there is a suggestion that the outer element should be changed twice before the inner. Let's get our values right once again and change them both—*unless* you are sure of the location of the air intake and the grain size of dust to which it has been exposed. If you are not sure, change them both.

Paper elements can be cleaned at a pinch. Remove them from the body, and knock them onto a piece of wood while rotating the unit. Knock the metal end in a rapid tapping motion. *Do not use an air line.* The air velocity is strong enough to drive the sharp dirt particles clear through the filter, making larger holes that allow large pieces of dirt to get through. Most good-quality elements can be washed (yes, in soapy, warm water). Wash them by shaking them gently under the soapy water and turning them around as you go. Once no more dirt seems to be coming off, rinse with clean water and place them in the sun on one end. After an hour or two, turn the element over on the other end, and it allow to dry some more. A quality Donaldson element is very expensive, and Donaldson will agree that you

Understanding and Checking Your Equipment

may wash them at least once. Buy the best and get the best service life. Buy cheap and pay dearly for it.

The steel element housing has a ¼" port on the engine side of the filter. This port *should* have a Purolator Indicator in it. This little black plastic indicator is a very important safety tool. Once the air cleaner begins to block, the vacuum on the engine side increases. When it reaches danger proportions, the little window on the Purolator Indicator shows a red signal which remains red even after the engine is shut down. This is to indicate that you must do something about the filter *now*.

Once you have attended to the filter, press the little button on the top of the indicator to reset it. If you get a chance, remove the indicator and suck on it with your mouth. You will be surprised at the little amount of vacuum required to trip it. The engine needs lots of *free* and *unrestricted* air.

The filter element sometimes has a set of small plastic fan blades at the inlet, designed to cause the air to swirl around as it enters the body and passes over the paper. The object is to throw large particles of dirt to the outside in the hope that they will drop to the bottom of the body and travel to the opening end. In the end cap is a rubber flap valve, like a duck's beak pointing downward. The dirt travels towards this and eventually falls into the rubber neck. When an engine is shut down, its last kick is normally backwards. This little back puff is normally enough to push the grit in the soft flap valve right out.

There is no point in having a good air filter if the air and dirt can get in through bad rubber boots downstream. Checking the system regularly and keeping clamps tight and properly positioned will help the engine have a long and happy life.

Batteries

The batteries on your engine are very important too. Without a good battery and easy starts, life can become a right pain in the rear. Batteries need to be serviced and looked after just like other parts of the engine.

The battery needs to be kept cool. For some reason, batteries almost always end up mounted right next to the exhaust pipe. Move them if you can. If you can't, at least get some heat-resisting product and make a deflector. Hot batteries discharge faster than cold ones. The electrolyte (acid and distilled water) inside the battery evaporates. No electrolyte equals no battery. And if you evaporate the acid away, no amount of water added will make the battery work again.

Batteries need to be insulated off their base. *Do not stand batteries on concrete or steel*, because they will discharge themselves through the casing. To prove this, remove the terminals, take a sensitive voltmeter, and put one pointer against the negative pole and another onto the chassis. You will get a voltage reading.

Keep the batteries clean and dry. You can wash them off with a hose, gently, but once washed, wipe them down to get the water off. Water makes contacts and causes batteries to run down. A dirty battery when wet will run your stored volts away to earth in no time.

Keep the battery terminals clean. They should not grow beards of powder; these are eating away your terminals and spoiling the contact between the terminal and the battery post. Wash these clean with a thick mixture of bicarbonate of soda and water (about two parts water to one bi-carb). Your wife may have some in the kitchen; steal a bit. Paint it on with a paintbrush, let it stand for five minutes and hose it off. Once the terminals are clean, coat them in thick grease to exclude the air and moisture.

When attempting to get a terminal off a battery post, *do not lever it off*. Use a screwdriver to force the clamp jaws apart, and lift it off. The terminal is cast onto a lead plate under the casing.

Understanding and Checking Your Equipment

This plate is about 3 mm thick, and it is easy to break the terminal off the plate. It is possible to have the battery repaired, but it costs almost as much as a new battery to get it done.

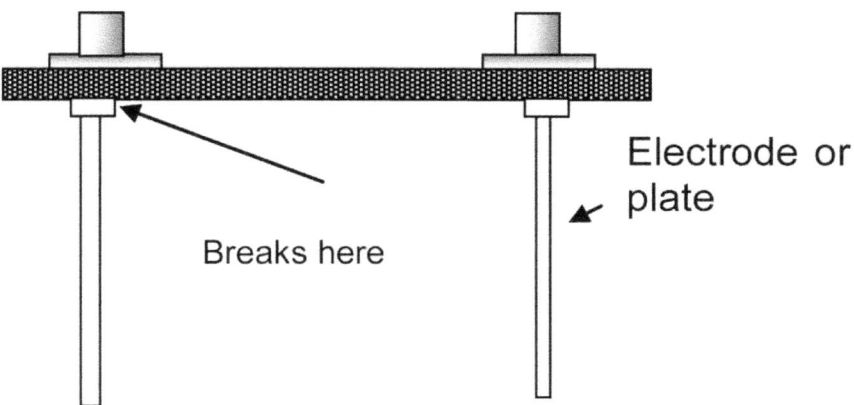

Figure 3.6. Cross section of battery

If you just crack the joint *and* you are using your battery hard, you may cause a serious explosion. The battery gives off a flammable gas, hydrogen, when charging hard or discharging fast. This gas builds up above the fluid level under the lid of the battery. Your loose contact can cause a spark—and *boom*, lots of bits of battery and acid spray everywhere.

The same applies to a battery being charged. Always remove the battery caps to allow for extra ventilation while batteries are charging. Always have the charger switched off while connecting or disconnecting the charger. A spark can set it off. If you have a dedicated battery charging area, keep welding machines away, and do not smoke there.

If you drain a battery for any reason, and if it was a good one, you can refill it. Take off all the caps, and hose it out with a garden hose and tap water. Turn it upside down to drain. Let it stand for a couple of hours. Your battery shop will let you have an acid water mix (electrolyte) to refill the battery. Fill it till the electrolyte is at least 10 mm over the plates, and place the battery on as slow a charge as you can. Fast charges heat up the battery, and too much heat can buckle the plates; if one plate touches another, it will cease to create energy. Each cell plate of a battery, normally six, provides 2 volts. If you damage one of the plates, it may provide only 1.5 volts or restrict the number of available amps.

Amperes, or amps, are what do the work. Volts on their own are harmless and gutless, but with a few amps they can start to provide energy. Each battery size is designated by the voltage—such as 12—and this is followed by a figure in amps per hour, or amp hour. When ordering batteries for an engine over 120 horsepower and particularly if the engine is direct coupled (without a clutch) to the pump, you will need a lot of amps to turn the engine over. You should ensure that the battery you buy is at least 200 amp hour or capable of 200 amps for one hour before it runs flat. You may be able to start your engine with a much smaller battery. But if you have to shut it down a couple of times in one hour, you may just run out of grunt and not be able to start your engine again. Two hundred amp hours should give you at least fifteen starts before it dies on you.

Your generator or alternator puts out about 14 amps. Your battery will charge at the rate of about 14 amps per hour till it reaches 200 amps. That is almost 14.5 hours of running to charge it from flat to full, assuming the battery terminals are clean, the battery is cool, the electrolyte is topped up, and the fan belts are tight. All of this just to start the engine—so take care of the battery.

Fan Belts

Fan belts must be kept tight. The belt that drives your fan, the long one, measured on the long side should have a deflection of about 6 mm if you push it firmly with your index finger with as much force as the weight of a brick, for example. The standard is 16 mm per metre of distance between the pulleys. They should not be guitar-string tight (they will destroy the fan and water pump bearings), but they should never, never be loose. If you hear a squeal when you start your engine, your belts are loose. Fix them now.

When replacing more than one fan belt, always order them as a "matched set of X no. B 54s", or whatever. Always order the belts as a *matched* set; belts with a common number always differ in length, and you will have one lazy belt if you don't.

Belts do not show signs of imminent collapse from the outside. Damage cracks form on the inside. The outer cover is the last bit to go. Check the inner surface when inspecting.

Belts come in various shapes and sizes, and all have a letter prefix indicating width followed by a number indicating length.

- A section is 13 mm wide.

- B section is 17 mm wide.

- C section is 22 mm wide.

There are a number of high-capacity belts that have different widths and thickness. In the main, industrial engine fan belts are A or B. Most have an engine part number stamped on them but are off a standard belt mould and can be replaced off the shelf for a third of the price from commercial drive-belt suppliers.

If your belt has shattered and you do not know the correct length to order, proceed as follows: Push the generator, or tensioning device, all the way in. Measure the outside diameter around the outside of the pulleys—*not* in the groove—with a tape measure. Tell your supplier the size in inches or millimetres. Also measure the width of the pulley between the faces to make sure you get A, B, or C belts. Also take the engine number off the engine plate so that, if all else fails, your supplier can look in the manufacturer's parts book and see if he can identify the correct belt number.

On some engines, the fan belt is toothed or has teeth on the inside edge of the belt to help it grip and to help it to bend around a small pulley. If you are taking off a toothed belt, replace it with the same. If you are in a pinch, you can get away with a smooth belt in place of a toothed one, but you must replace it as soon as possible.

Turbochargers

The cost of turbocharger repairs due mostly to operator error makes it essential that we talk about them.

The turbo is a compressor provided to force air (oxygen) into the cylinder under some pressure to assist the diesel to burn hotter and better. This causes an increase in power of as much as 25 per cent. The exhaust gas discharging out of the engine is expanding and, given the added push of the piston on the exhaust stroke, is racing out of the manifold. The turbo uses this gas flow to drive the compressor. The turbo can be clearly seen; it is mounted on the exhaust manifold just above the engine and looks like two large, round doughnuts mounted side by side.

The unit is made up of two sets of turbine blades, slightly cupped and delicately shaped to provide the maximum flow with the minimum disruption and turbulence. The blades are mounted onto a common shaft that is supported in the centre by some very delicate high-speed bearings between the blades. The exhaust gas coming out of the engine drives one set of blades, and the other set drives clean air into the engine. The turbo rotates at speeds of up to 7,000 rpm once it gets wound up.

For these high-speed bearings to survive in the heat and at the speed at which they travel; they need a constant, unrestricted,

Understanding and Checking Your Equipment

and clean oil supply. The oil pump pumps this supply from the engine sump up through a series of drillings and into the bearings. The used oil simply drains back to the sump.

If you shut the engine off at full speed, you cut off the bearing's oil supply—bearings that are spinning at 7,000 rpm. They will continue to spin for at least the next three or more minutes, gradually slowing to a stop, all *without* any oil. I understand that Volvo make a stationary diesel engine with an oil bath around the bearings; this is the only engine that can be shut down without a slowdown period. Check with your supplier.

Most of our pumps are fitted with remote *emergency stop* shutdown switches. Most operators use these to stop the engine under normal operation. *Wrong.* That is damaging to the turbo. In an emergency, it's fine to damage the turbo; injury to an operator is not acceptable. That is the only time an emergency stop button should be used. A turbo overhaul costs about 3,500 dollars. Do you really want to spend that amount of money twice a year? Do not give me the "I've been doing it that way for years" bull. Have you still got a turbo that is working? How bad is your fuel consumption? Do you blow clouds of smoke when you start up? Lost some power lately? Engine getting tired?

The correct shutdown procedure is to bring the engine to a slow idle for at least two minutes and then shut it off.

I understand that this is a problem with direct coupled pumps, automatic shutdowns, and remote controls. If you have these items fitted, look for a normally aspirated engine or have a system where a clutch is fitted and the pump shutdown controls can be fitted to the clutch. A torque converter lends itself well to this application.

The purpose of a turbo is to provide the engine with more oxygen so it can better consume its fuel and therefore provide more power or provide the same power with less fuel. Turbocharged engines have specially set fuel injectors and have slightly different timing settings. Whichever way, if you have gone to the trouble and expense to provide a turbocharger, why is your engine running at half revs and half power? Is your engine working hard? If it is not, you are abusing it in one way or another: crankcase dilution is thinning your oil; heavy carbon is building up around your turbo blades, piston rings, and valves; *or unburned fuel is burning in your turbo.*

Fuel burning in your turbo is burning off the sharp edges of your turbine blades, rapidly reducing their efficiency, laying deposits of coke onto the portings, and overheating the already straining bearings. Any uneven deposits on the turbine blades unbalance the turbo. Can you imagine what an imbalance could do at 7,000 rpm? Your pistons could end up with lots of little bits of turbine blade embedded in them.

When your engine is shut down, it almost always stops with at least one piston with both valves open (between exhaust and inlet stroke, a V8 may have two sets open). On the top of the exhaust is a flap valve fitted to keep the rain out. This valve flaps open when the engine runs and closes when it does not.

If you fit this flap to the exhaust so that it faces forward when the truck is traveling along the road, the flap will open and act as a scoop. Air will be driven down the exhaust, through the turbo, through the valves, and out through the air cleaner. The finely balanced turbo needs little encouragement to spin; it will start to spin and can build up a fair old speed *without* any oil. After a couple of hours of this, the bearings are gone, and your engine will not run. Can you imagine how much SSG is inside your engine now? This problem also exists for normally aspirated engines; the dirt can do serious damage when you eventually start up. *Think.*

Turbocharged engines need to be serviced very regularly, and the oil must always be in good order. A very clean and regularly serviced air cleaner is particularly important with a turbocharged engine, as the turbine can create a vacuum and draw oil from the turbine bearing area through the oil seals on either side and squirt pure oil into the cylinder. Oil will not compress.

The engine will not run without the turbo. Take care of your engine, and consider the poor old turbo.

Cold Starts
Before we leave the engine, it is important that we discuss cold starts. If you are liable to get into frosty or very cold climates, it is advisable to fit heater plugs to your engine. These can be fitted to most engines by the injectors to heat the fuel injection area. Another neat idea is to fit one into your sump drain plug fitting; this keeps the oil, and the engine, warm.

Never use proprietary brands of ether, or whatever, to start an engine. You will do major damage to the crankshaft; you will blow a gasket; you will damage a piston; and you will most likely destroy the exhaust. By spraying this rubbish into your engine, you are putting the engine under strain it's not designed for. The resulting explosion of the ether plus diesel places unknown strain on the piston, which will likely fire before it reaches TDC and try to fire backwards. Remember the knocking noise? That's what that is. The engine is turning clockwise by the starter, and it fires before the piston gets over the top; the resulting anticlockwise strain on the crank can only be imagined.

If you do not have heater plugs, carry one of those little 20-dollar butane gas torches around with you. Heat up the inlet manifold at the point where the air cleaner joins the manifold, where the manifold branches. Heat it just warm enough so that you can just hold it. You should not burn the paint. Turn the key, and she's away.

Under normal running conditions, no engine needs to be assisted to start. If it does, it's a very sick engine and needs some TLC ASAP. Engines respond to love and care just like humans.

Prestart Inspections

Please try to remember that breakdown time is *downtime* and that downtime costs more than the direct costs you can see. It impinges directly on the company budget and on the marketing plan. Downtime can ruin a good company. Your job security depends on eliminating downtime.

It follows that you must look after your tools and equipment; that is what you are paid for, and that is what keeps you employed. To ensure that you do this, better employers provide you with inspection and maintenance checklists. As previously stated, these lists are made up not because you are thick but to jog your memory and to ensure that, in the heat of the moment, you do not forget to check the oil or whatever.

Another smart thing to do is to run a logbook. Get a page-a-day diary and write down everything you do on a job. Add thoughts and notes about "nice if I had", "gotta have", etc. This helps you to remember what you did so you can repeat it, gives you a shopping list, and reminds you to get more O-rings or whatever before the next job. A quick look in the log before leaving in the morning could save you a great deal of drama and coming back to the yard to pick up some forgotten nozzles or a hose joiner.

Look after your equipment, and get to know it really well. This is what puts food on your table and pays the school fees. Also, your employer is paying you a compliment by letting you take charge of almost half a million dollars' worth of equipment; he most likely hocked his home to be able to buy it for you to use. Respect that commitment.

Understanding and Checking Your Equipment

DAILY MAINTENANCE CHECKLIST

ACTION	OK?	Sign
Pump engine oil level checked		
Fuel tank full checked		
Pump oil level checked		
Radiator water level checked		
Radiator clean checked		
Batteries fluid level checked		
Gear box oil level checked		
Fan belts checked (spare set?)		
Air cleaner tell-tale checked		
Rubber hose boots on inlet okay?		
Drive belts checked (if fitted)		
Coupling checked		
Clutch play checked		
Mounting bolts checked		
Filter(s) cleaned (spare bag/element?)		
Batteries level checked (pump engine)		
Batteries terminals clean		
Truck fan belts checked		
Truck engine oil checked		
Truck fuel level checked adequate		
Truck engine oil level checked		
Tyres/wheel nuts inspected		
Spare tyre, spanner, and jack checked		
Lights and indicators checked		
Brake fluid level checked		
Emergency shutdown operational		
All gauges in good order		
Rego, in date checked		
Windscreen clean, wipers checked		
Mirrors in good order checked		

Figure 3.7. Daily maintenance checklist

High Pressure Water Jetting – An Operator's Manual

EQUIPMENT CHECKLIST

ITEM	Qty	Sign
Supply hose x 20 metre lengths, good couplings		
Hydrant/hose adaptor (fees paid?)		
Hydrant valve handle		
Hose ramps		
HP hose, inspected and in date		
HP hose O-rings		
HP hose joiners		
HP hose restraints		
Hand gun inspected and in good order		
Nozzles for gun		
Pipe bombs suitable for task, threads in good order		
Hose guides suitable for task		
Remote shutoff valve/switch, or		
Foot valve		
Barrier tape and signs		
Face shields		
Mono goggles		
Gloves		
Wets		
Metatarsal protectors		
Rubber boots		
Ear muffs and plugs		
First aid kit—complete		
Drinking water		
Hand cleaner		
Wipe rags		
Manhole lifter		

Understanding and Checking Your Equipment

Clump hammer		
Broom		
Shovel		
Mechanical Tools		
Rope		
HAZIDs and Work Instruction		

Figure 3.8. Equipment checklist *before* **leaving base**

Chapter Five

Understanding Drains and Pipes

Drains and pipes—sounds simple, right? Just round things with a hole through the middle through which water, sewerage, or whatever travels, right? Nothing could be further from the truth. These are very complicated structures, and if we are to do our job properly, we need to know more about them.

Let's look at the different styles of round things with holes and discuss, in simple terms, how we are going to deal with them.

The pressures mentioned for concrete pipe work are for direct jet striking the concrete. Smart operators can place their jets anywhere in a pipe with surgical precision, and these guys can use higher pressures. The rest of us had better keep our pressures down for now.

Most concrete is porous—that is, there are microscopic gaps between the bits of sand and stone used to make it. The water jet seeks out these little porosities and uses them to get into the material and break it up. The smaller-diameter pipes that have been spun have less porosity than the bigger-diameter ones and will be able to absorb more pressure without damage. We can do little or no damage to nonporous surfaces with a water jet.

CONCRETE PIPE—max jet pressure 150 bar (2,000 psi)

This is a common sewer and water carrier and comes in all sizes, from about 250 mm up to as much as 2 metres or more in diameter. The larger ones are normally gravity or fall lines and are not often pumped or under pressure.

The pipe is normally made in a mould with two parts, between which is fitted a "drum" of reinforcing steel rod like a net. The gap is filled with concrete that typically has ¼" aggregate (stones) in it. Some of the really big pipes may have as large as ¾". In the smaller sizes, this concrete mix is then spun at great speed to force the air out and to close all the gaps. In the larger sizes, the mould is vibrated to get the air out. This spinning leaves a very compacted and smooth finish on both inside and outside surfaces.

Normally the reinforcing is located as near to the middle of the section as possible, with at least 25 mm of concrete between the water or product and the reinforcing steel mesh.

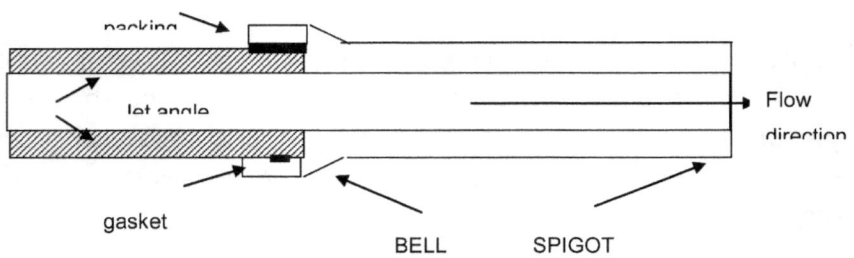

Figure 4.1. Bell and spigoted concrete pipe

It is normally "belled and spigoted", with one end like a bell (female), which surrounds the spigoted end (male) of the next pipe. Normally, like our water jet fittings, the flow direction is from the male into the female. As in the sketch above, the flow direction is from left to right.

The gap between the bell and the spigot is fairly loose; on larger pipes, there could be as much as a 50 mm difference. This area is packed with molten lead, oakum (like rope hammered in), or cement. In some there is a rubber gasket not unlike a large oil seal. If the seal or packing is not correctly fitted or has come loose, the fitting will leak and allow tree roots in.

When cleaning with a water jet, the back shooting nozzle must always be in the same direction as the flow. In most cases, this is no problem, as we always clean pulling the spoil out, downhill, in the direction of the fall (direction of the flow). If we run the nozzle in the other direction, uphill, we stand a very good chance of blowing the packing or the seal out, causing the pipe to leak and allowing roots to get in.

All concrete sewer pipes suffer from damage due to hydrogen sulphide gas, given off by the decomposing sewerage. No sewer pipes ever run full; they are normally designed to run three-quarters full with an air gap on top. The gas floats on top of the fluid and "eats" the exposed or unwetted concrete. The gas attacks the lime in the cement used to make the mortar, and the cement "rots". This allows the aggregate to fall out, exposing the reinforcing, which rusts and drops into the flow. Eventually this concrete cancer makes a hole right through the pipe. The pipe then has to relined or replaced.

I often wonder why some clever person has not made a spray gun that could be dragged through the pipe after installation, sealing the top half with something H_2S will not eat?

It follows that when we are cleaning these pipes, we minimize the jet impact onto the top quarter or so of the pipe surface. This surface is always loose or rotten, and a 100 m section incorrectly cleaned could result in five tonnes of sand and aggregate being removed unnecessarily. The client will be a very unhappy chap. If the pipe is to be relined, the customer will want the entire surface cleaned of degraded concrete. Be aware that this will result in a very large quantity of solids that will have to be removed from the manhole, so allow for removal and disposal in your bid price—normally four or five times the cost of cleaning.

Get in the manhole (*after* gas testing and using the appropriate confined-space procedures), and mirror up the pipe to try to ascertain at what level the pipe has been running at—quarter, half, three-quarters full—then decide on your nozzle configuration. Do you want the nozzle to run along the bottom of the pipe or run higher, requiring a skid or centraliser? The idea is to remove the offending deposits on the floor and sides but to minimise the jet impact onto the roof.

NOTE: It is a bit tricky to look up a sewer pipe unless you are happy to fill your ear and hair with nasties. Sewer pipes are always on the floor of the manhole, so we use mirrors to see what is happening. "Mirroring" is the judicious placement of mirrors that allows you to see long distances up pipes and sewers. You need two good-quality truck mirrors—the kind that mount and pivot from one end. Mount one onto the end of an adjustable roller-painting extension handle (aluminium, available from most paint shops), and fit it with a firmly mounted clamp.

Lower the extension bar unit into the bottom of the manhole, and use the other mirror to reflect the sun's rays directly onto the bottom mirror and angle it so that you can see up the pipe. Beats a video hands down. If you can get into the hole (confined space entry, guys; watch out; harness on with tag line tied off), hold the bottom mirror with your hand while your standby reflects the sun onto it from outside.

If you pull back a lot of aggregate, a gasket, or, heaven forbid, some reinforcing, stop cleaning and make a note of where it came from and tell the client before proceeding further. He may require a video inspection to be carried out to check the extent of the degradation. If you carry on cleaning, the roof might fall in.

So, what nozzles do you use? The most important question is, What are you trying to remove? What is the reason for you being called out? Is the pipe blocked? Is there a reduced flow rate or backing up in peak periods? You need to know these things.

Problem: Pipe Restricted or Blocked
What causes a flow restriction or total blockage in a sewer line?

A bundle of disposable nappies in the line should reasonably travel, but they may be hung up on something. But what? First we need to get the blockage out. We would use one forward and maybe three back-facing jets at 30 degrees.

OR

Roots have a tendency to run in the direction of the flow along the pipe, and a big tree could have several meters of root in a pipe. If roots have been there a long time, they could block the pipe right up. Alternatively, they could act as a stop and hold up those disposable nappies. Use a root cutter nozzle that would possibly have two side at 70 degrees and three pushers at 45 degrees.

OR

A gasket incorrectly inserted can end up half in the pipe, forming a stirrup into which fibrous items have hung up. The root cutter would most likely get it.

OR

A section of pipe diameter is lying inside the pipe across the bottom quarter, which is not uncommon in new pipelines.

When cutting a spigot down to fit into the line, pipe fitters often break the removed ring, and a piece of it can end up inside the pipe once it is closed up. This is sloppy workmanship, but it does happen. When the pipe was new, it most likely serviced a developing area, and the flow rate was minimal. The problem was not noticed then, but as development took place, the flow rate increased and the gap inside the pipe became too small for the designed flow rate due to the blockage.

This can be a problem to identify, because the nozzle may travel under or past this kind of blockage and give no indication of its presence. A large amount of fibrous product or plastic bags will come out though. This should be an indication that something solid is holding up the spoil; getting that loose stuff out has not solved the problem. Similarly, bricks from manhole construction work can be left inside the pipe.

How do you get it out? Build a fishing tool as below.

Step 1: Get a road sweeper bristle. These spring steel bristles from sweeper brushes are about 2 mm thick by 4 or 5 mm wide. These fall out, and sometimes the operators keep them.

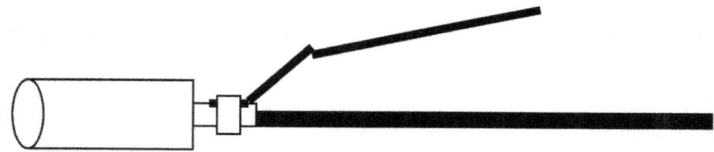

Figure 4.2. Fishing tool

Step 2: Clamp three or four of these "fingers", about 250 to 300 mm long, to the hose ferule immediately behind the bomb or nozzle holder. Use a substantial hose clamp—the bolted kind. Bend them in a vice, carefully; they might snap, as above.

Step 3: Insert the nozzle into the pipe, and let it ride up fairly quickly until it clears the obstruction. Slowly pull it back; you will feel the "fingers" lock on. Gently ease the load back out of the pipe. If you get hung up, a really good yank on the hoses will free the fingers from whatever they are hung up on. Do not use heavy-duty bar or rod for the fingers; they will get hung up, and you will be unable to get your puller out.

Problem: Protruding Laterals
Plumbers often punch a hole in a sewer line and insert a piece of PVC pipe to carry the waste from a building. This is supposed to be flush with the inside of the sewer pipe wall, but I have seen some that are halfway across the diameter. These will hang up a plastic shopping bag with ease, creating a "parachute" that fills up with solids, eventually blocking the pipe. A plastic shopping bag will travel all the way to the sewerage farm unless it hangs up on something. If you get one as described above, find out why or what it is hung up on and get it fixed (see "Laterals" below). If you don't, you will be back. To get the blockage out, use one forward nozzle and three back as above.

Problem: Pipeline Collapse
This too is not uncommon, particularly where building work has been taking place. A big earthmover operator, unaware that there is a sewer below, has applied a heavy load on the pipe, and it has fractured. This can be a total collapse or just a "green stick" fracture. I have pictures of star pickets and piles driven clear through a sewer line.

During your cleaning, if you get a piece of pipe wall coming back or a load of soil that should not be in the pipe, stop and recover your nozzle, get a video, and take a look. Many operators have experienced the problem of the nozzle escaping out of a pipe through a break or crack and have either lost the nozzle and hose permanently or had the nozzle surface in a pavement some distance away, causing all manner of complaints from the public. Wash enough outside soil back, and the pavement or road will collapse, causing all kinds of unwanted drama.

Problem: Reinforcing Collapse
Hydrogen sulphide has rotted away the cement, and the reinforcing has dropped into the flow path. This is a very sick sewer but not uncommon. How do you tell? The best way of course is to run a video up it, but you do not often have access to one. Or you could mirror it; 20 to 30 metres up, you can see considerable detail if it is a bright, sunny day. Primarily you are going to get a lot of aggregate back. If you are not pulling it off the top of the pipe with your jet, it has fallen down during normal sewerage operation. You will undoubtedly have trouble getting the nozzle back on your return run. It goes forward easily enough but stops at the same place on pullback. When it hangs up, the hang-up is not solid but springy; it feels like someone has put a big elastic strap onto it. That is normally reo bending and springing back.

If you do get a hang-up like that and cannot get the nozzle back, try to run it up to the next manhole, disconnect it, and pull the hose back. If all else fails, tie a rope on the hose and pull it back with the truck—a bit drastic, but you have to do what you have to do. Make a note of the distance from the manhole to the problem and report the problem to the client. Inspect your hose carefully before using it again.

EARTHENWARE SEWER OR DRAIN PIPE—150 bar unfired; 300 bar fired
The standard length of this type of pipe is about one metre long. It is made from a clay material not unlike a brick. In some cases, it is used unfired, which means that it has not been hardened by heating it. This is easily recognisable: it is light red-orange, chalky, and dull. Fired earthenware looks like a fired house brick; its surface is shiny and in places blackened by the heat. High pressure can be applied to fired pipe but

not to the unfired. The nozzle must be kept moving at all times in unfired pipe; if it stops for any reason, get the pressure off ASAP.

Earthenware pipes are always laid in the ground with the spigot discharging into the bell. The bell and spigot is normally sealed with oakum or greased hemp hammered in; in the old days it was commonly filled with molten lead. It can be cleaned only from the bottom up, with jets pointing back. It suffers badly from roots entry.

CEMENT-LINED STEEL OR IRON PIPE—max jet pressure 250 bar (3500 psi)
NOTE: When cleaning at 250 bar over concrete, the nozzle must be centralised and kept moving. Do not stop travel while the nozzle is under pressure, or damage to the lining will result. If working in the bottom of the pipe, do not exceed 150 bar; add flow rather than pressure.

This style of pipe and lining is not as common for sewerage as all-concrete belled and spigoted pipe, but it is often used if the line is under a bit of pressure. Lined pipe is more common in water supply lines. This style of pipe is also found plastic-coated for water service but mostly cement lined for sewerage.

The style of join has a great deal to do with the system pressure. For low-pressure supply, the pipe *could* be cast or spun iron with bell and spigot ends. It could be butt jointed end to parallel end with a bolted split or tapered clamp around it. The clamp contains seals that fit around the outside of both ends, forming the seal. There is normally a small gap between the ends to allow for expansion and contraction of the pipe due to external weather conditions. Be careful when water jetting these pipes; you can blow the seals out with a well-directed water jet.

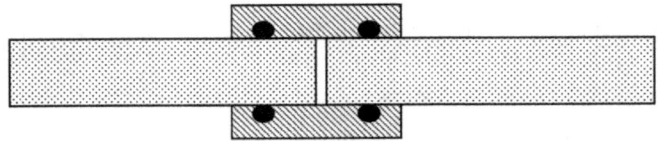

Figure 4.3. Typical slip clamp joint

Alternatively, the pipe can be welded end to end, or flanges could be welded or cast on and the flanges bolted face to face.

Cement lining is "spun" in on the smaller lines. A section of pipe is set up on rollers, cement is pumped in, and the pipe is spun at high speed using centrifugal force to press the cement onto the walls and to produce a smooth, even wall thickness. On the bigger pipes the cement is troweled on with a machine that sprays the cement onto the walls, and a rotating trowel comes along behind and mechanically trowels the cement flat.

Typically, the cement lining is 12 to 20 mm thick. There is no steel reinforcing in the lining. Once cured, it is very hard and abrasion-resistant. However, the bond to the pipe wall is often not good. The strength of the lining is in its structure. Being a perfect ring, it is very difficult to break.

Some linings contain small pieces of stainless wire or glass strands as a strengthening material. This is nasty stuff, as the wire can cut you and irritate your skin. Wear leather gloves when working with reinforced linings. They are very likely to damage your hoses, so allow for this in your bid price. These are very common in flue linings carrying hot air or gasses.

On welded or flanged lined pipe, the internal join point is coated after assembly. This is done either by putting a man up the pipe with a bucket of mortar and a trowel or mechanically with special tools. Once sealed, the join cannot easily be seen from the inside on video inspection. These pipes can be cleaned in either direction, as there are no pitfalls or gasket seals to worry about.

Problem: Pipeline Flow Restriction or Blockage

Blockage due to silting is common in areas where a long pipeline tilts downwards through a dip in the land or under a road, and heavy particles have settled there. This is not likely if the flow in the line is constant and of a reasonable velocity. Where the flow is intermittent, solids can settle at low spots. Use six or eight back, preferably alternate vee and pin jets.

The lining may be cracked and broken out. This is a common fault, particularly where the pipe has had an impact or been bent. A section of the lining can fall out and jam itself in the line. This can act as a trap for silt, and a wall can rapidly build up, eventually blocking the flow. This is very difficult to remove; even though you can get the silt out, the chunk remains. Try the spring "fingers" bomb we discussed for cement pipe.

Another cause of lining failure is water getting behind the lining, rusting the pipe. The rust buildup bulges up under the lining, pushing it off. When we discuss nozzles and tools, we will discuss lining removal in more detail.

As the joints are sealed, there is little likelihood of a gasket or misalignment causing a blockage; however, keep your wits about you and think. If the pipe is on the surface, take a walk along it and look for possible problems. I once had one running through a farm. The farmer used the pipe as a surface to hammer a plough shear flat after bending it. Using the pipe as an anvil, he broke the lining on the inside. Who, me?

One of the more common jobs on this style of pipe is total lining removal so that the client can recoat the inside with a new lining. This is a difficult but not impossible operation. We will discuss how under nozzle diameter selection later.

PVC OR PLASTIC PIPES—maximum jet pressure up to 690 bar (10,000 psi)

PVC pipe is nonporous. It has no little holes in it for the water jet to get into, and so quite high pressures can be used. The nozzle should not be directed at the same point for a long time, as erosion of the pipe wall may take place.

PVC pipe is a very common product nowadays, but it is normally not used above about 150 mm (6") in diameter.

PVC pipes are normally belled and spigoted and are laid like concrete sewer pipe, with the spigot running into the bell. In most cases, these joints are glued together and the bond is fairly strong. However, a directed water jet can get between the two layers and peel them apart. So, try to ascertain the lie of the bell and the spigot, and run the nozzle accordingly.

For very low pressures, these pipes are sometimes joined with a push-in coupling bell-and-spigot joint with one or two seals inside. The coupling can move easily, and buried pipe can become disconnected. Try to find out the joint style before you start work. With this style of joint, drop your pressure to about 300 bar, and never go against the flow direction; the gaskets come out with very little pressure if struck by the jet. Due to their extremely smooth and nonporous surface, these pipes do not often become fouled or clogged.

Problem: PVC Pipeline Flow Restriction or Blockage

Crushed pipe can be caused by clumsy backfill, compaction—either intentional or accidental—or vehicle traffic. It may sound illogical, but a perfectly round pipe passes more water than the same pipe made oval. The velocity of the product going through the pipe at the crush point is also greatly increased, which may wear out the pipe at that point.

Understanding Drains and Pipes

If you push a nozzle up the pipe and it stops suddenly, the chance of a crush being the cause is very likely. Not a great deal can be done about that from the inside. Mark off your hose, lay it the same distance over the ground, and mark the spot. Tell the client, and let him dig it out and replace the crushed section.

A broken pipe or a separated bell and spigot are also common. Plumbers sometimes forget or incorrectly glue a section. When the backfill takes place, the joint moves and comes apart. If the flow is blocked or muddy or if there is a nice green patch on the grass nearby, you can be fairly sure the line is broken or separated. Do not let the nozzle run; it could be outside the pipe, and you may never get it back, or it might come up through the ground in an embarrassing place. Mark the approximate position on the ground so the client can find the problem.

Plastic bags and similar objects can block the pipe simply by slow flow; the bag sits in the line and silt gathers around it until it is wedged in. A good flush with a fire hose will often shift these. Alternatively, use a drain nozzle with one forward shooter and three or four back. Use lots of flow and low pressure.

Scale, cement, grease, and so on can stick to the walls and may become a problem. Cement is a particular problem where PVC pipe is used as a cable conduit through a concrete structure such as a bridge. During the pour, someone steps on the pipe and pulls out a bell-and-spigot joint, and the cement gets in. Not too much of a problem for an FR (forward spinner with side cutters) combination nozzle, as high pressures can be used in these circumstances. If you start getting bits of plastic back with your wash-out, be careful that you have not run out of the pipe; it might be a good idea to restart from the other end.

BRICK SEWERS AND DRAINS—100 bar jet pressure maximum

There are not many of these left (maybe older sections of Sydney, Melbourne, and Wellington built by convicts). Normally, brick sewers are pear shaped with the wider section at the top. In 90 per cent of cases, the cement has long been dissolved, and the bricks are not bonded. They are not easy to clean with a water jet, as the bricks fall out and cause even more problems. Ideally, they should be cleaned by hand; in many cases, they are big enough for a man with a fire hose to get through.

Blockage problems are as for all other types of sewers. Discuss using the water jet and its implications carefully with your client before working. Make him aware of the damage the jet might do.

There are other styles of pipe that you may come across; one very old style is made of brown paper and glue. Do not get involved with that one. Of course there are also the common screwed and threaded steel pipes; these are not usually used for sewerage but are very common for water. The problems with these are rust and water-deposited scales. Both will come out reasonably easily, as you can use higher pressures here. A chain flail is good *if* the client intends to coat the line when you have finished. You will remove sections of galvanizing, and without a coating, the pipe will rust dramatically and rapidly.

PIPE LININGS

Many modern pipes are internally coated with cement, plastic, or thick paint. At one time or another, you are going to be asked to get them out so the client can recoat them.

As mentioned above, cement linings are normally either spun on or mechanically troweled on. These linings can be extremely strong, although they are not well bonded to the pipe walls.

Understanding Drains and Pipes

Their strength is in the fact that they are a "ring" around the pipe. Unless you can break the ring, you will not remove it economically.

Ideally you need to set up a nozzle square onto the surface, drag it through the pipe, and cut a slot in the ring. Once it is cut into two half rounds, getting the pieces out is relatively easy.

Make a tool as below in figure 4.4. Use about 75 mm diameter, stainless bar stock for the body. Drill holes for the water. Set the nozzle standoff from the pipe wall at about fifteen times nozzle diameter. Run a bomb up the pipe to the exit point, remove the bomb, and fit the slitter.

As you apply pressure to the nozzle, the tiller arm will kick up and align it square to the lining. Pull it back slowly without twisting the hose. Remove the slitter and refit bomb; run the bomb back up, and refit the slitter upside down. Do not rotate the hose, simply repeat the pullback.

NOTE: HP hoses do not twist. The writing on the outside of the cover is normally straight along the hose; where it is on the top at one end, it will be on the top at the other. Use the writing as a guide to tell you where your cutting nozzle is. If it starts to turn, you can give it a twist to get it back straight again.

Pull back slowly; the slower you go, the more chance of completely severing the lining. Patience here will save hours later.

You should now have two pieces of half round cement rattling around inside the pipe. Get a good flushing bomb (four holes at 30 degrees), and work the lining out. One thousand bar should be enough pressure to smash it up; use as much water as possible.

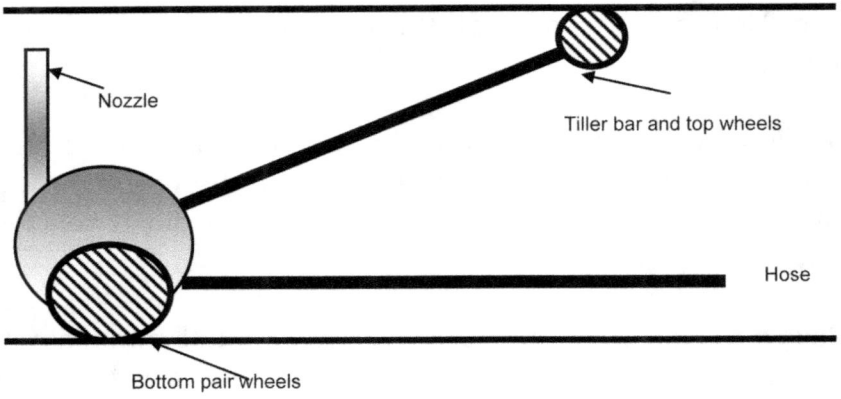

Figure 4.4. Lining slitting tool

Plastic Linings

This is a loose term for a number of "flexible" linings placed inside a pipe. These vary from epoxy tar (a black, two-part bituminous product), straight epoxy resin (clear to about weak tea), and two-part plastic that is mixed at the nozzle and sprayed onto the pipe walls (typically blue or yellow).

The most common material that you are likely to encounter over the next few years is the epoxy tar coating. It is no longer used, and there is some indication that the leaching chemical is carcinogenic (causes cancer) and is being replaced with one of the two-part plastic inert products now available.

Epoxy tar can be chipped off by impact; a rattle gun will break it off easily, but it is a bit of a trick getting a rattle gun up inside a 6" pipe. A pressure of 1400 bar will shift it, but if the water is heating up due to friction, it can melt the product and cause it to stick somewhere else, which will drive you potty. Add a tiny amount of dish-washing liquid to your water tank.

My approach with this product is to use a Rotofan drive at about 20 rpm (the bigger the pipe diameter, the slower you go) with a two-hole nozzle (one side at 90 degrees and the other back at 45 degrees).

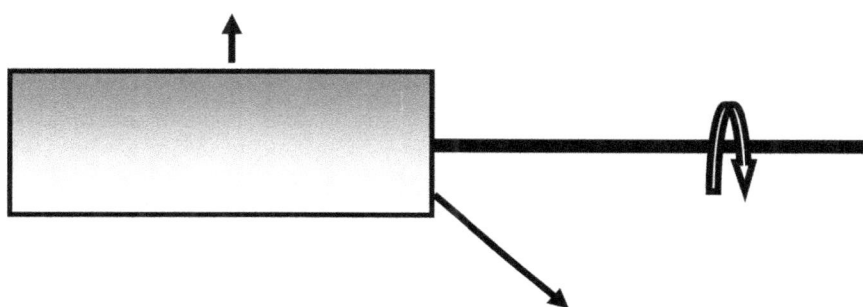

Figure 4.5. Rotofan "bomb" nozzle

The concept of a Rotofan nozzle is—the side or 90-degree nozzle pushes the holder up against the pipe wall, allowing the 45-degree nozzle to attack the deposit from up close. In this manner, a 50 mm bomb can clean 600 mm pipes without the need for skids and nozzle extension tubes.

The side nozzle is always slightly larger than the back nozzle to make sure that the back nozzle stays firmly against the pipe wall. As you engage the hose-turning machine, the side nozzle points to the centre and the back nozzle points at the wall of the pipe.

The selection of nozzle combinations requires you to think a little about what you are doing. You should draw a sketch and calculate the nozzle selection reactive force. If you have 100 kg of reaction on the 45-degree back and only 20 kg of reaction on the 90-degree side, the nozzle is going to turn and come back out of the pipe. By doing the calculations, drawing a sketch, and drawing in the force lines, you can see what will happen. This way you can be sure that the bomb is going to do what you want it to do.

Reaction = 0.0224 x litres per minute x √Bar pressure = kg

A 50 mm diameter Rotofan nozzle can be used in all pipe sizes, right up to a couple of metres in diameter. The side jet slams the bomb against the pipe wall and the 45-degree does the cleaning.

Run this nozzle up and back, and follow it with a chain flail nozzle.

Figure 4.6. Rotofan hose-turning machine (manufactured by T-Squared)

Water enters through the big pipe on the lower right, passes through the swivel, centre left, and out through the swivel shaft to the hose. The swivel is rotated by an air motor and 30:1 and 100:1 gear boxes.

Chain Flail

The ideal chain to use is a second-hand transmission or sprocket drive chain; it is cheap, has rough ends, and will take quite a beating. Any decent-sized rotating nozzle can be modified into a chain flail.

Understanding Drains and Pipes

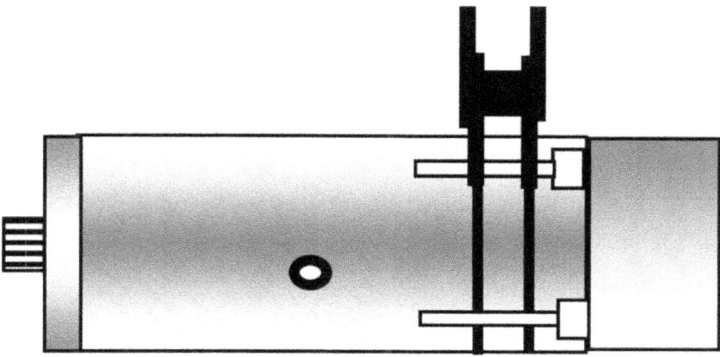

Figure 4.7. Sketch of chain flail concept

Cut two grooves around the body for the chain links to fit into, drill upwards from the base, and thread in a long machine screw, which will pass through the link holes. Make the chains about two links longer than the pipe radius, so they drag around the wall.

If you like, make your own 1000 bar chain flail as above with two pushing nozzles and two turning nozzles. Make the body out of 630 stainless and the rotating barrel out of aluminium bronze. The clearance between central shaft, end caps, and rotating body needs to be no more than 0.03 mm. That is the difference between the inside diameter of the hole and the outside diameter of the shaft. This needs some really good machining to get that small a tolerance.

You might, if you are clever enough, machine three or four labyrinth seal grooves inside the barrel on either side of the nozzle outlet holes. These need to be about 1.5 mm wide by 2 mm deep. The drive nozzle should point back at about 25 degrees and should be drilled offset to centre by 6 mm. If you cannot make your own, there are several makes on the market, including DCS Engineering in Melbourne, Saloteck in Holland. and ENZ in Switzerland.

Two passes with each will most likely get the lot. If it doesn't, you might have to run the Rotofan one more time.

If you are having trouble with hot water (friction somewhere?), and epoxy or tar is re-bonding after you have removed it, add a little dish-washing liquid to your water pump's header tank.

The ideal product is Shell Teepol, which is biodegradable and environmentally friendly. In a screw-top, 1 litre, plastic drink bottle, punch a 2 or 3 mm hole in the bottom and another alongside the cap. Fill it with Teepol, tie a piece of string around the neck, and drop it into the water tank under the inlet. It will *not* harm your pump. If you get soapy foam, you are adding too much. Get another bottle and make the holes smaller.

On average 1 litre should last about four hours. This is what is known as a wetting agent and will prevent the tar, epoxy, or grease from re-sticking onto the wet walls. Use this system when removing fat from sewers too. Keeps your hoses, wets, and face shield clean too.

Epoxy, plastic, and fiberglass can be a real test of your skills. You need a sharp chisel and medium hammer—for example, 1500 bar and 60 litres per minute. Once again, I would use the Rotofan to do this, running at about 100 rpm. If you do not have one, use a hose that you can turn through 180 degrees and back by hand. Fit a three-hole nozzle at about 75 degrees with really good nozzles, and you should get it. Hard work though. A 6" pipe could take about one hour to do 6 metres. A lot depends on how well bonded the product is. This job can then be finished with the chain flail, which will leave a good surface onto which the client can bond the new lining.

Removal pressures can be calculated. Get a portion of the product, send it to a soils laboratory, and ask for a crush-test figure. This information is normally available in psi, and you double it for the pump's pressure. If it has a crush of 6,000 psi (400 bar), use 12,000 psi (800 bar) to get it off. The same applies for any product you wish to remove with a water jet.

All water jetted steel surfaces will rust within minutes of cleaning. If you need to stop rust bloom forming, spray the surface with a rust inhibitor as soon as you have finished. Talk to your local chemical supplier, and he will come up with something suitable. I have dripped it into the water-jet header tank as the pump was running; it did not harm the pump and did the job quite well. You use only a little bit, and it is not expensive. Rust inhibitor leaves a gun-metal black coating on the pipe.

> **IMPORTANT NOTE**
> When working a bomb up a pipe *always* first check that the bomb and hose ferule is longer than the diameter of the pipe being cleaned. If it is not, the bomb can turn around in the pipe and come back out at you—and possibly kill you. Fit a short length of suitable tube between the hose end and the bomb to create the required length. I once had one turn on me at 600 bar. Luckily, it missed me, but it hit a railway sleeper and split it. At 600 bar, the water is leaving the nozzle at almost the speed of sound—1,440 klm/hr. A bomb running freely can reach highly dangerous speeds.

LATERALS
Laterals are smaller service connections into a main line. Typically a lateral will be inserted to discharge the waste from a private home into the main sewer line. These are normally PVC and normally not bigger than 100 mm (4").

Builders are supposed to cut off the lateral so that the insertion does not enter, or is flush with, the internal bore of the main pipeline. They also are supposed to "make good" the entry point to prevent leakage from the sewer line and to prevent the ingress of roots through the hole. They often do not.

They are supposed to use circular cutting tools, giving the PVC insert a tight fit, but usually they punch a hole with a hammer and chisel. This often breaks chunks off the pipe wall, which fall inside—another problem for you. This damage often exposes the reinforcing of concrete pipes, and once the reo starts to rust, major damage at this point is guaranteed.

If the plumber has not "made good", roots will enter at the lateral insertion point. As they grow, they apply pressure in the hole and can split the concrete pipe.

A number of tools can be used to cut off laterals. First, you have to find them. This means a video survey needs to be carried out and the insertion point found. Dimensions are then taken—xx metres from manhole # yy North—then you need to identify the material: PVC (typical), earthenware, or metal (it is unlikely to be metal).

Cutting off a PVC lateral can be done mechanically with a rigid drain-cleaning kit attachment. These have a rotary saw that can cut ahead of itself.

A water jet driven circular saw cutter is also available out of Switzerland and the United States (look up water jetting/drain cleaning tools on the Internet). These are adequate for roots, plastic, and unfired earthenware but really struggle with anything harder. However, you can do the job at a price.

Grit-entrained water jet cutting systems do a good job but are costly. Water alone will not cut PVC. It will cut unfired earthenware but takes a while. Fired earthenware is glass hard, and water alone is unlikely to get started, let alone cut it.

The pin jet nozzle (at least 600 bar) needs to be firmly skid mounted on an arm that can be externally rotated around the wall of the main sewer pipe. It should point fractionally inward, so that there is no chance of the jet stream hitting the main sewer pipe wall. You will need video support to do this so that you can see what you are doing.

Understanding Drains and Pipes

Normally, removal is done by locating the lateral, inside the pipe, by video, marking the point on the surface, and then digging down until finding the joint point and removing the offending pipe.

ROOTS
Another "make your own":

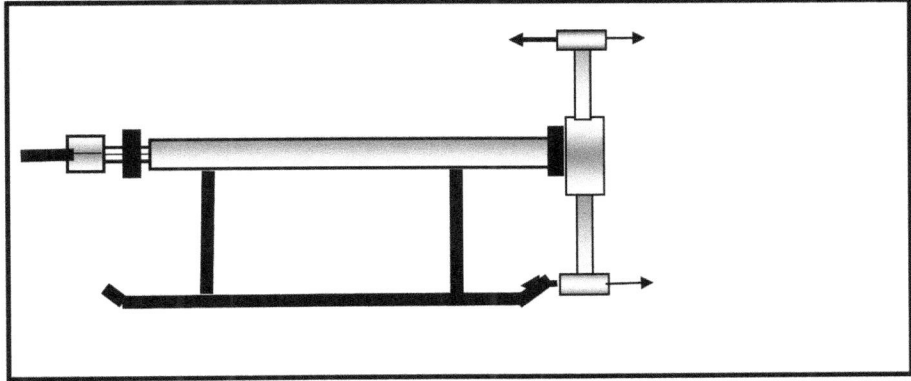

Figure 4.8. A sketch of a simple root cutter you can make

A centre pressure pipe (1/2") attached to the hose at one end and to a T piece at the other is fitted snugly inside another piece of pipe (1") to which three or four skids are welded to suit the size of the pipe being cleared.

The centre pipe is fitted with two big washers welded onto each end to contain the pressure pipe inside the big pipe and to give the tee and hose fitting something to wear on.

Into the tee, fit two arms or smaller pipes, maybe 3/8", and fit a tee onto the end of each piece, such as 3/8" x ¼" street tees. Into the ¼" tees, we typically fit ¼ MEG Spray Systems pin jet nozzles. One pair faces forward and the other back. The nozzles facing back will be larger than the nozzles facing forward. This encourages the skid to travel up the pipe until it reaches the blockage. You might consider making one arm shorter than the other so the jet stream strikes a different piece of the root.

All pipes and fittings used must be rated for at least 2.5 times the maximum working pressure set up on the relief valve fitted to the pump. For all pressures over 300 bar, the pipe or tube must be seamless. The pipe may burst along a welded seam.

The operator then twists a loop into his hose and turns the arms backwards and forwards, clockwise and anti-clockwise around the pipe wall, cutting off the roots close to the pipe wall. When cleared, the unit will travel forward on its own until it hits the next root. The back-facing jets will wash the cut off roots back and out.

This is a great tool for earthenware, as the nozzles are set parallel to the pipe walls and do not hit the pipe walls, and so will not damage them.

Chapter Six

What Nozzle Holder (Bomb) to Use?

Nozzle holder selection is critical to efficient sewer cleaning. Primarily you need some detailed information. But how do you get it? Build a job query sheet; you most likely have something similar already in Operations. Without this information, your job is much more difficult to do:

Example of a Job Query Sheet

Question	Answer
Where is it	John Street
The pipe diameter	450 mm
Distance between entry points	80 m
What it is made of	concrete
What system of jointing	bell and spigot
How sealed	gasket
How old is it	17 years
What runs through it	sewerage
What the client thinks the problem is: blocked, restricted flow, periodic back up, or leaking	blocked
0. Between where and where	Manhole 10–13
1. Depth from surface	3 metres
2. Direction of flow	from NE to SW

Figure 5.1. A typical simple job query sheet

What do we know (or think we know) from the above? A certain amount of local knowledge is essential.

1. John Street is mainly old, residential houses but has a fish-and-chip shop and a deli on it. There are several trees along the pipe route. The blockage will most likely be made up of roots or domestic waste and/or (also likely) fat from the two shops.

2. We will most likely need a centraliser to get the nozzle up into the middle of the pipe. If there is fat in there, we need to get amongst it, not under it. The centraliser will need to fit into a 450 mm diameter pipe, with a six-hole nozzle at about 60 degrees.

3. We will need 80 metres of jetting hose, plus the depth: total 83 metres.

4. Being a reasonably old concrete pipe, it will have some degradation in the air gap due to H_2S, so we will need to run at low pressure, about 150 bar, with lots of water. We might have a lot of aggregate to get out, so we might take a gully sucker or vac-truck to lift it out. Getting a possible cubic meter of stone out by hand is no easy task. If we are taking a sucker, should we block off the downstream pipe? Take an inflatable plug or baffle plate.

5. The jointing method means that we can wash only in the flow direction: downhill (entry from SW end). There is a distinct possibility that the spoil will run quickly past us and not settle in the manhole. A method of blocking off the downstream is now essential (see 4 above).

6. The client tells us that it is blocked, so there is likely to be a head of spoil above the blockage. We will need to open the upstream manhole (no. 10) to see what the fluid level is. If it is high, then our downstream working manhole (no.

What Nozzle Holder (Bomb) to Use?

13) may well overflow when the blockage is breached (pollution). Can the sucker keep up with the flow? Perhaps we should not block the downstream pipe, but once the blockage is clear, clean out several downstream sections after the blockage has gone down the line.

7. A 3- m deep manhole will require us to work the hose from the surface, as entry may be dangerous should the blockage clear suddenly. We will need tied-off, full-body harnesses when working over the manhole. If we are going to enter the manhole to insert a plug to block off the downstream line, we will need to adopt Confined Space Entry procedures and have had training.

See how much information you can glean out of a few simple questions? Asking the questions makes you appear professional to the client and makes your job easier.

My selection for the above job would be:

- 100 litres per minute at about 150 bar (2,000 psi) (R=27.5 kg)

- 6-hole retro nozzle about 50 mm diameter, with 1.5 mm nozzles at 60 degrees and 30 degrees off the hose line

- Four back and one forward 50 mm nozzle with 1.2 mm forward and 1.6 mm back at 30 degrees to break through blockage if required

- Two-hole retro about 35 mm diameter with 2.8 mm nozzles at 45 degrees to the hose line for roots

- 100 metres of ¾" drain (thermoplastic) hose (0.3 kg/metre = 90 metre travel)

- Take a centralizing skid, but try to run in the bottom first

- Mirror set

- Gas tester (see discussion below)

- Safety harness and tether (see discussion below)

- Inflatable pipe plug 450 mm or Weir board and sandbags

- 20-litre can of dish-washing liquid. (Put ½ cup into water tank every minute or so to keep fats from sticking back onto pipe wall and on your hoses. The water in the drain should not foam. If it does, reduce the amount of soap you are putting into the tank.)

Gas tester discussion. Once the manhole has been opened and barricades fitted, lower the tester into the hole and test for gas present. Check at floor or fluid level, and then test half a metre above the floor. Most sewer gases are heavier than air and may well be lying on the floor, waiting for you to stir them up. Be aware that there may be no gas about *until* you start water jetting. When you stir up a pile of still sewerage and other waste, gas will be released. So keep checking; if possible have the tester around your neck while working. It will be a good idea to open manhole covers above and below your work site. Ensure that these are secured so no one, especially children, can fall in while you are not there. Do not forget road traffic signs and cones.

Safety harness and tether discussion. Any job where you risk a fall because you are working over a hole, or whatever, should require you to wear a safety harness with the tether tied off to something solid, typically the truck. Rising gas could render you unconscious, and you might fall into the hole. Similarly, a small trip or slip, and you are in. Tie yourself off with the tether as short as is practicable. If you are deliberately going to enter the hole, you *must* be formally trained and certified for Con-

What Nozzle Holder (Bomb) to Use?

fined Space Entry and have all the crew and equipment present and erected.

NOTE: Using harnesses—

- You must carefully inspect your harness before you fit it. Check the webbing and all the stitching. If in doubt, don't use it.

- You *must* be trained to fit a harness properly; incorrectly fitted harnesses can cause very painful and possibly permanent injuries to your genitals.

- Keep your tether as short as practicable; limit the distance you can fall.

- If you do fall, who is going to pull you out? You should not be expected to hang in a harness more than eight minutes; if you do, the resulting blood clots could kill you. Always plan a rescue before starting work. One man alone will most likely be unable to lift you. Always consider "what if".

- Never use a safety belt; these are *note* designed as fall-arrest devices. Crushed kidneys can result. Always wear a full-body safety harness when a fall is possible.

Chapter Seven

Nozzles and Holders

What is a nozzle? For lack of a better word, a nozzle is the spout which forms the shape of the jet of water. At the same time, it creates the water jet velocity by creating a restriction onto the water flow. We can replace a nozzle, but we cannot replace a jet. The jet of water is formed by a nozzle. We do not use pressure to do our work; we use water weight and velocity (speed the water it travelling). We work with pressure inside our hose and nozzle; once it leaves the nozzle it has no pressure, only mass and velocity.

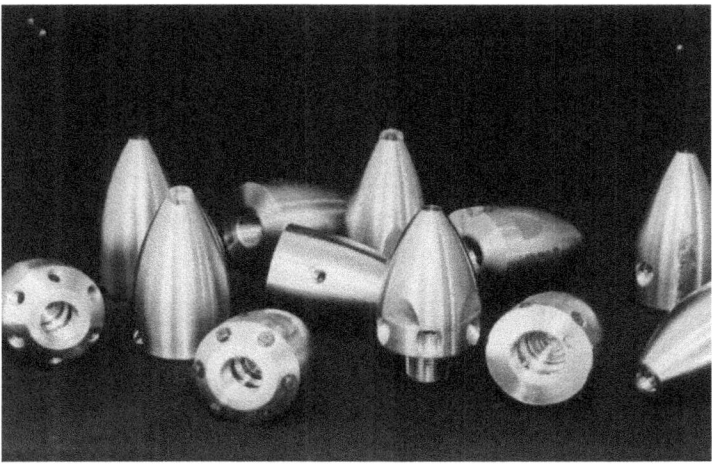

Figure 6.1: A selection of nozzle holders for various duties (manufactured by T-Squared)

There are two types of nozzle shape: pin (straight) jet or vee (fan) jet. Both are good tools if selected with some thought. The vee jet is a good pusher and acts like a shovel or rake. Do you have lots of soft loose product? Select a vee jet. Hard, lumpy stuff? Select a pin jet to break it up, and then use a vee to pull it out.

When ordering vee jets for a nozzle installation, order 15-degree vees, which are available from 15 to 90 degrees in most sizes. A bigger angle than 15 degrees distributes the same amount of water over a bigger area, and all that's likely to happen is wet. Big angles are great for washing down with a gun, but for real work, keep the fans tight.

Figure 6.2: Vee and pin jet (1/4 MEG from Spray Systems)

A comment you will often read from this point on is one of the prime considerations when planning any water jetting job. Please consider it and store it away:

> *Consider pressure as the sharpness of your chisel blade and consider flow as the hammer you hit it with.*

If you have a very sharp chisel and no hammer, you will cut very fine grooves into your surface but are not likely to blow much away. If you have a reasonably sharp chisel and a medium hammer, you will cut into the medium and break chunks of hard stuff off. If you have a damn great hammer and a blunt chisel, you will blast away great chunks of soft stuff, with waste flying in all directions.

Consider a flying matchbox and a flying half brick, in relation to flow. Which will do the most damage: a matchbox travelling at 100 km per hour or a half brick at the same speed?

Big water streams will do more work than small water streams, so we need to get as much water through individual nozzles as is practical for the job—biggest possible hammer or half brick. At the same time, we need enough chisel blade to get through the stuff in the first place, so a balance is needed. The balance should be as much on the side of the hammer or half brick as possible.

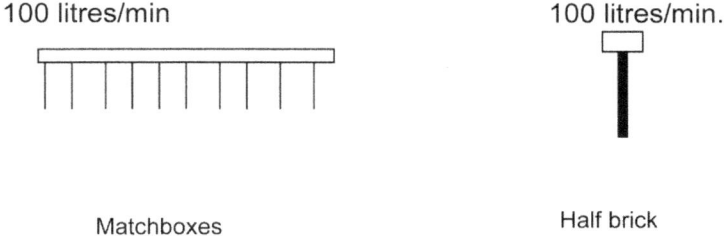

**Figure 6.3: An indication of
10 nozzles = matchboxes and 1 nozzle = half brick**

I too often hear operators saying, "If only I had more pressure". With a little bit of thought, you may well reduce the pressure, double the flow, and halve the job time. Remember: volume then pressure. Think of flow and pressure in these terms, and you will progress nicely.

In some applications, the sharp chisel is needed to penetrate into the substance. Once penetrated, however, you need a decent hammer to blow it away. You need high flow and high pressure. For this reason I understand WOMA has recently created a nozzle with two jet streams, one from each of two pumps: one ultra-high minimum flow and the other low pressure with high flow. The small one cuts the product up, and the big one blows it away. It's complicated and expensive, but productive and "just what the operator needed" as we become more technical and professional.

Some applications require a diamond cutter to cut fine materials, and the sharp blade on its own is enough (high pressure and very little hammer or flow). It is typically used for eggshell scale in heat exchangers; packed, thick roots; resins; and hard scales in process pipe work.

Some applications require a huge flow or hammer and very little chisel point, such as mud and silt removal from pipelines—like washing down with a fire hose. So we need to decide whether we concentrate our water stream out of 1, 2, 4, 6, 8, or 12 holes. What does this job need? Think about the chisel and hammer—big bullet, little bullet—and decide how many holes you need in your bomb to do the job you have to do. Twelve holes will shift lots of sand but will not cut roots and vice versa. Consider a garden rake. Lots of tines move lots of little items; with two tines, you move nothing but you do tear up the ground.

Having decided on the chisel and hammer balance, we have a pipe-cleaning job that we have decided needs a sharp blade and a heavy hammer. We have roots and solids in the pipe, so we look into our toolbox and pull out a 12-hole retro. *Wrong*. That may be a sharp blade (if it is new) with a little hammer (12 little hammers). We dig a little further and find a 4-hole retro. That's better, though 2 holes might be nicer. A 12-hole retro is lots of sharp chisels with no hammer; we might do better work

Nozzles and Holders

using a fire hose or shower rose. The more holes you have, the more individual little squirts of water you are throwing at it. Remember the matchbox and the half brick? A 12-hole is a washing tool for bringing back sand and soft mud.

Think of a big bullet and a little bullet. Both travel at a speed of 2,000 metres per second, for example. The little bullet has a slug weighing 1 gram, and the big bullet has a slug weighing 100 grams. Which will do the most damage? (Which one would you use to shoot an elephant?) The big bullet, of course. It has weight plus velocity and hammer plus a little sharpness; the little bullet has a good sharp blade but little hammer. The little bullet is a mass of small nozzles. The big bullet is one big one.

To come to some understanding of what pressures to use, we might take a sample of the material to be removed and have it tested for compression strength. This figure will be given to you in psi or kg/c2. It is the load applied to the material before it collapsed, This will indicate what water pressure will be needed to cause it to break. To be sure you have enough multiply the compression factor by 2, and you have an indication of the water pressure required to penetrate and remove it. These tests can be carried out cheaply at a concrete-testing or soils-testing laboratory. The sample needs to be as thick as possible.

Having decided to develop a new way of looking at flow and pressure, let us also take a look at our vocabulary.

We do not *cut* material when we are cleaning with a water jet. We may penetrate it, we may crush it, but we do not cut it. Cutting is done by specific tooling and in most cases includes an abrasive in the jet stream. So we blast it away. We do not cut roots; we shatter them or blast them apart.

What normally happens is the water jet is directed at a product; the water enters small, almost microscopic, holes or pores in the product; the water in the pores is pressurised by more water trying to get in; and the pore bursts.

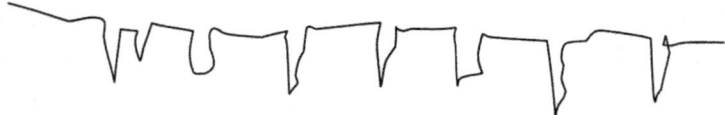

Figure 6.4. Porosity magnified

If you were to place a section of wood, concrete, or rock under a microscope, the surface would look something like the sketch above. What to the naked eye looks like a smooth surface is filled with holes and craters.

As the nozzle travels over the surface, it pressurises the craters and bursts them away, exposing another set of craters below and so on. These holes are termed "porosity". If the product has no porosity (such as granite, PVC, resins, and steel), we cannot remove it with just plain old water at average pressure. We need to add an abrasive to "punch" holes into it (or to cut it) or an ultra-high pressure pump (see compression test above).

As we think about porosity and pressurising the hole, we must consider something called "dwell time"—that is, the time the jet spends over the hole, pumping it up till it bursts.

Typically, spinning nozzles—or screamers, as they are sometimes called—are the major culprits when considering lack of dwell time (the time the jet remains over one spot before moving on) when trying to remove hard product. Because of their speed, they are often of very little use. These tools have two or more holes drilled into a tubular body, mounted onto a fixed shaft attached to the hose. The water squirts out of the holes, causing the body to spin faster and faster until the water stream is a blurred sheet of fog.

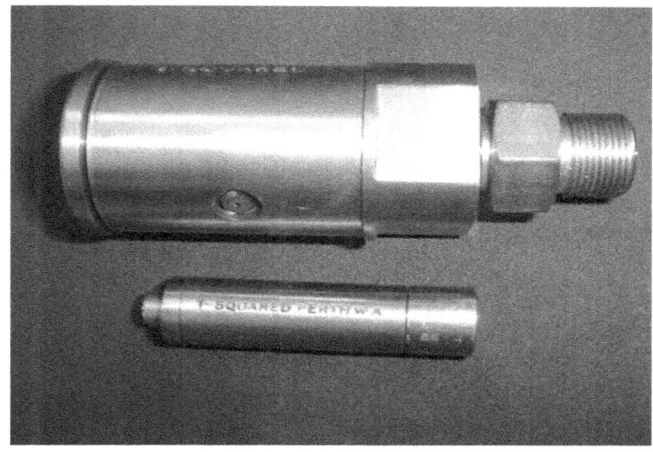

Figure 6.5. Spinning nozzles, 12 mm and 50 mm (manufactured by T-Squared)

Let's take a look at the average self-propelled spinning nozzle. This unit typically rotates at speeds up to 3,000 rpm. Some manufacturers fit a "breaking" nozzle next to the driving nozzles in an attempt to slow them down.

In my opinion, this does not really work. The nozzle accelerates as long as the spin nozzle reaction exceeds the break nozzle reaction.

As a hypothetical example of this speed, imagine that we have the pump running on the back of our truck, and the truck is travelling at about 140 klm/hour. Holding the gun fitted with a pin jet out of the truck window, we point the gun at the road surface and pull the trigger. What is happening?

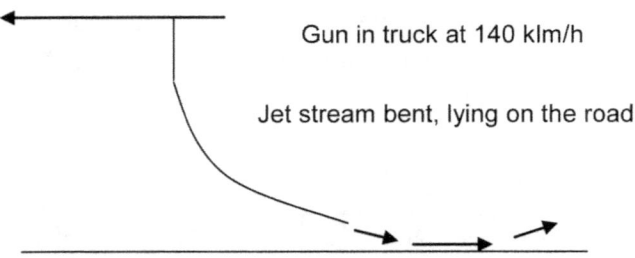

Two things are happening: the jet is bending and being laid on the road, and before it has a chance to create any damage, it has moved off to the next point. Being laid at such a steep angle, it is most likely bouncing off the surface back into the air again. Dwell time equals 0.

As we slow the truck down, the jet straightens up and begins to damage the road until we come to a stop, at which time it drills a hole through the road. Dwell time equals 100 per cent.

So, dwell time is a very important factor to consider. How long is enough? The only way to find out is by trial and error. Listen to your nozzle operating up the pipe. With hard scales, you will hear what my crew called "Rice Krispies snap, crackle, and pop". When the dwell time is just right, you can hear the product bursting and splitting. No noise change equals no activity.

A fairly new innovation for the spinning nozzle is to offset the barrel to the shaft. As it spins, it tends to vibrate. This vibration knocks the nozzle body against hard scale and cracks it off.

I recently made a big one (100 mm, 4" diameter, weighing 7 kg) fitted with tungsten carbide teeth for use in the alumina industry for attachment to a 280-litre-per-minute, 250-bar pump. Video footage of it working showed excellent performance rates.

Nozzles and Holders

Figure 6.6. Vibro nozzle

The tool pulverised the scale. One pair of back-firing nozzles (also acting as brakes, just visible below the rim) helped push the scale back and out as well as driving it 10 metres up the pipe pulling a 1" hose

Another use for the spinner is the chain flail discussed earlier in chapter four.

How fast do you let the hose travel up the pipe? It depends on the product being removed, but you *must* think about it. What is being removed? How hard is it? Is it porous or smooth? What are we doing: are we breaking it up so we can pull it back on the return run, or is it already loose and just needs to be liquidised and flushed out?

How much product do we want to remove per pass, or more realistically, how much product *can* we shift? And, if applicable, how fast do we go to remove spoil but not damage the pipe? This is something that cannot be detailed in a manual but requires you, the man on the hose, to think about. Think about it you must, as it is a very serious consideration, such as when trying to remove bonded scale. "Any old retro bomb" is not good enough. As a professional, you must be able to look at the job, think a little, and choose the required bomb.

If you have only one, you are in trouble and need to seriously consider getting more. I would recommend as many as six variations as a minimum stock on your truck.

Consider having the following:

DUTY	PUSHING	CUTTTING	AHEAD
Roots	2 at 30° off hose line	2 at 80° off hose line	None
Roots and nappies	2 at 30° off hose line	2 at 60° off hose line	One ahead 5°
Sewers/ with solids	4 at 45° off hose line	None	None
Sewers/ general duty	2 at 30° off hose line	4 at 60° off hose line	None
Hard scale	3 at 60° off hose line	None	None
Grease and fats	8 at 45° off hose line	None	None
Sands and mud	8 at 30° off hose line alternate vee & pin	None	None

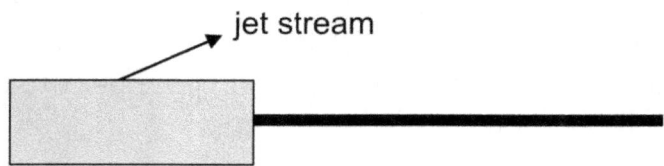

Figure 6.7. Sketch explaining the angle

Nozzles and Holders

You will most likely be wondering about the 5-degree ahead nozzle? The placement of forward shooters exactly in the middle of the bomb with the jet straight ahead is a pain. By offsetting it slightly, the nozzle will drill a hole ahead bigger than the bomb is; if straight, it will drill a little hole right through but not big enough to allow the bomb through to allow the backshooters to do their job. The 5-degree offset will cause the bomb to travel up the pipe a little cockeyed, but that is good; by rotating the hose, the scale is attacked at a multitude of angles.

Do *not* forget the extension tube to prevent the bomb turning around inside the pipe. *If* you want the bomb to travel exactly straight, drill the forward shooter exactly opposite one of the backshooters, and make that backshooter one size bigger than its mates. Test it inside a pipe—*not on open ground*. Wrapping a hose with a bomb on the end around your feet is no joke. An interesting thing to do is to get a clear polycarbonate pipe, locate your nozzle in it, and see how it performs.

STANDOFF AND ANGLE OF ATTACK

The next thing to consider is the nozzle standoff and angle of attack. Pipe bombs come in all angles of water jet.

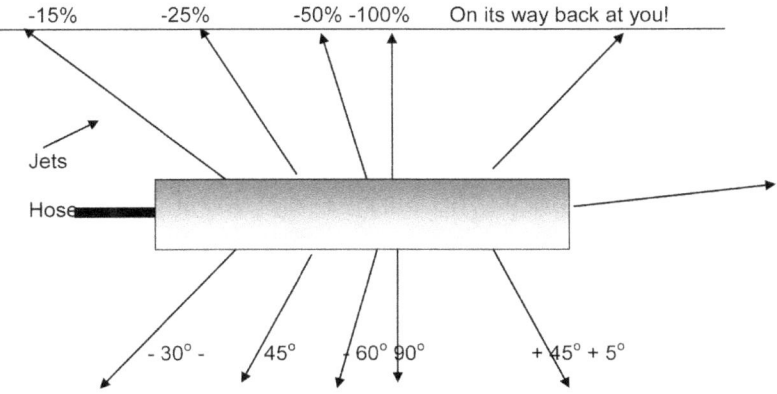

Figure 6.8. Various nozzle angles available "off the shelf"

The angles sketched above are typical of standard bomb nozzle inserts. You can purchase off-the-shelf bombs with nozzles at those angles or combinations of angle. We lose a certain amount to the nozzle angle, so subtract from your calculations to allow for the angle.

Why so many angles? Imagine that you are the pipe bomb with the hose attached to your feet. Place a nozzle in each hand and raise your arms straight out on either side of you. Start the imaginary pump, and allow the water to flow out of the imaginary nozzles. What is happening? You are blowing a hole in the wall on either side of you, but as to travel along the pipe—nothing.

Now, if you lower your arms towards the ground, you will gradually begin to "lift off" until your arms are pointing towards the floor. At that time, you will start to travel up into the air. It follows that the maximum *reactive*, or pulling, force exerted on the hose is when the nozzles are pointing straight back along the hose line.

Now go stand in a doorway and look at how the distance between the nozzle outlet and the door frame changes as you lower your arms. With your arms at 90 degrees to your body, the nozzles are touching the doorframe. A great breaking tool and a good jet form penetrates the product and breaks it up. However, there is no "lift" on your feet, no hose travel up the pipe, and no "washout" effect.

As you lower your arms, the lift or hose travel begins to happen. Look carefully at your "nozzle" fingers, and you will see that the water now needs to travel a great distance before it hits the doorframe. Has the jet stream dissipated or died before it reaches the pipe wall? The ideal standoff for steel pipes with hard scale is 10 to 15 times the nozzle diameter. A 1.5 nozzle should be not much more that 25 mm away from the product being removed.

Nozzles and Holders

We might consider two pushing nozzles pointing back at 30 degrees off the hose line for travel with two sides at 85 degrees cutting product off the wall of the pipe.

You need to decide what angles or combination of angles you need and get those onto the truck so you can do your job properly. When deciding on a nozzle configuration, do yourself a favour and make a scaled sketch of the job, place your nozzle into the sketch, and extend the jet lines. Scary? Time to build another nozzle holder? You need to watch your angle for two reasons: available reactive force and angle and distance of attack (standoff) onto the pipe wall.

Reactive Force can be calculated using this formula:

$$R \text{ (kg)} = 0.0227 \times \text{ltrs/min} \times \sqrt{\text{bar}}$$

- 0.0227 is a factor

- ltrs/min is the flow rate of your pump (or the nozzles you intend to use)

- $\sqrt{\text{bar}}$ is the square root of the bar pressure. Look at most calculators, and you will see a button with the square root emblem $\sqrt{}$. Enter the pressure in bar, such as 1000, and press the $\sqrt{}$ button. The number that comes up will be 31.62.

So, for a pump delivering 1000 bar and 80 litres per minute, the reactive force will be

0.0227 x 80 x 31.62 = 57.42 kg of force acting on the nozzle

For a pump operating at 300 bar and 180 litres per minute the sum will be

$$0.0227 \times 180 \times 17.32 = 70.76 \text{ kg of force}$$

Easy? Now it gets a little more complicated. Say the nozzles are set at 45 degrees on the above bomb; we must subtract 25 per cent (-25 and hit the % button). The force on the hose is now 53.07 kilograms.

Do not concern yourself with the number of holes or jets at this stage. Except for a nominal friction loss, the number of nozzles has little bearing on the reactive force.

What can we do with these numbers? For the pump above, if we want to push a nozzle vertically up a drain pipe on the side of a building, how far will the nozzle travel up? The ¾" hose weighs 1.6 kg/metre add water 0.5 kg/meter = 2.1 kg/metre.

We know we have a total force available of 53.07 kg, divide by 2.1 = 25.27 metres straight up. Useful stuff to know.

Cleaning from the bottom up is much smarter that from the top down. As long as you do not have a sea anchor for a nozzle holder, ignore the weight of the bomb for now.

Normally, running horizontally in a perfectly smooth and *clean* pipe, a hose weighs about 20 per cent of its weight in air, so the hose will "drag" about 0.4 kg/metre. We now divide our reactive force of 53.07 kg by 0.4 = 132.67 metres along the cleaned pipe. If we know that the manholes are 100 metres apart, we can get there with this pump and this hose. *However*, this is possible only if the hose is traveling on a clean pipe floor and not lying on top of (and settling into) disturbed mud. Clean a heavily fouled pipe in 3 m segments—up three, back three, up six, back six, and so on. The bomb can shift only about 150 to 200 kg of suspended spoil at a pass. Do you want your bomb to travel through a long pipe? That's easy if it is clean most of the way.

It follows that the more "back" the angle is, the further we can travel. *But* the further back the angle is, the greater the distance between the nozzle outlet and the pipe wall. This is called standoff. Standoff is a very important factor when working on hard scales or removing concrete or plastic linings.

For maximum impact and penetration for hard scale removal, we need a standoff of *no more* than 10 to 20 nozzle diameters. A 1.2 mm nozzle (1.2 x 20 = 24) 24 mm standoff is the distance between the nozzle outlet and the wall of the pipe. The steeper the nozzle angle, the greater the distance the jet must travel before it reaches the pipe wall. At the same time, as the nozzle angle steepens, the chance that it will simply bounce off the surface increases. So we need to keep it as straight on as we can.

Selecting a Bomb Shape
When selecting a bomb nozzle set shape for a pipe-cleaning job, you need to ask yourself certain questions.

1. What is the pipe made of?

2. What diameter is it?

3. What is in the pipe that needs to be removed, and is it well bonded like scale or loose like sand?

4. Located how?

5. Do I need a big hammer and sharp chisel, or what combination of flow and pressure do I need?

6. Lots of jets or just a few?

7. How far up the pipe do I need to go? (manhole to manhole or flange to flange)

8. How much product am I likely to draw back? (This matters in a manhole but is not important if the pipeline discharges to floor.)

9. What is my flow and pressure?

10. What is the weight per metre of my hose?

Let's try to plan a bomb for a sewer job. Here are the answers to the questions above:

1. Concrete, in good condition, used for storm water

2. 450 mm

3. Road dirt, sand, and soil

4. Lying on bottom third

5. Big hammer, not much chisel

6. Medium, 6

7. 90 metres

8. A simple way: Measure the distance across the top of the spoil—say that is 350 mm. Now measure the height from the bottom to the top—say 150. The area of a triangle is the base 350 x ½ the height (150 ÷2 = 75). It is easier to break this down into fractions of a metre to get a result in cubic metres: 0.350 x 0.075 = 0.026 cubic metres per metre x 90 metres for the pipe length = 2.36 cubic metres, or almost 5 tonnes of spoil coming out. If this is suspended sand and mud, you can plan on removing about 20 tonnes of spoil. This calculation misses about 15 per cent of the spoil as the triangle's sides are straight where the pipe wall is curved, but it will prepare you for a lot of work to come. Very few people ever work this out, and then they get a big surprise when they start to clear away what has been brought back with the bomb. It can be worked out accurately using mathematics but the above is easier when all you have is a simple calculator and a ruler. If you are bidding the job and need to factor in transport *and* disposal, add about 30 per cent

Nozzles and Holders

to your total bid price. If you are also removing the pump water, the sum becomes massive.

9. 300 bar at 100 litres per minute (R=0.0224 x 100 x 17.32 = 39 kg)

10. 0.25 kg/metre (R divided by 0.25 = 155 metres—heaps)

Back to our bomb. This product is reasonably soft and not bonded together, so all we need to do is to get under it, lift it, and get it moving back to the manhole. We need an aluminium bomb about half the depth of the spoil, such as 75 mm diameter. To sink it into the mud, we fit a small forward shooter set at about 5 degrees off center to push the nose down. Six back nozzles should do the trick. We want lots of push/pull power, so we will set them at about 30 degrees off the hose line.

What nozzle holes should we use? I suggest 10 litres (1 mm) out of the front jet, leaving 90 litres divided by 6 = 15 litres per minute each (1.2 mm) out of the back. *See nozzle selection pages 106/107.*

In a clean pipe, 6 x 15 litres = R 34.91 kg—1 x 10 litres 3.9 kg = R31.01 kg forward = 124 metres forward travel.

How far up do we travel with each pass? A quick sketch helps to explain why we need to think about this. The hatched lines indicate the spoil in the pipe.

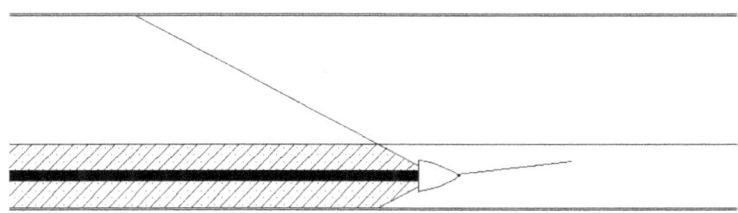

Figure 6.9. Sketch of hose and bomb in spoil

High Pressure Water Jetting – An Operator's Manual

Think a little: how much mud and sand can the nozzle carry back? The picture shows 1 metre of soil behind the nozzle. The jets are going to pick it up, fluidise it, and start it moving back. Scaled in a 450 mm diameter pipe, the hatched section is 1 metre long. The fluidised material will remain fluid and move for how far before it stops moving and settles solid again? One metre, 2 metres, or 4 metres? We have about ¼ cubic metre (1/2 tonne) of soil for every metre.

Four x ¼ m^3 metres is 2 tonne in weight or 1 cubic metre in volume. We have it up and moving, and so we pull our nozzle back slowly, pulling it all the way out. We will most likely move about 1 cubic metres of soil *and water* with each pull if we are very lucky.

There is absolutely no point in trying to go all the way up 90 metres and then pull back if all we are going to shift each pass is 4 metres worth of spoil. As we come back, we will drop what we have and pick up more until the pipe is still a third full at the far end.

At the same time, we are going to disturb the surface, and the hose is going to settle into the spoil. The spoil will grip the hose, preventing it going forward and making it very difficult to pull it back. So the operation is up three or four, back three or four, up six, back six, and so on. It is time consuming, but you will end up with a clean pipe a lot quicker. (See 8 in the list above.)

Now, the front nozzle or jet is pushing the tip down. What happens if it turns over? It will push the bomb out of the soil. This is an option too; if the nose lifts, the back jets are blowing back and down. When fitting the bomb to the hose, check out the lines of writing, which give the working pressure details.

The water jet hoses do not twist. When attaching your bomb, if you look at the relationship of the forward nozzle and the writing, and fit the bomb in such a way as the writing is on top while the nozzle points up, and keep the hose traveling with the writing up, all will be well. Want to go up the right hand branch of a tee? Point the offset front nozzle towards the left hole; the bomb will neatly slide around the right corner. Works every time.

The success or failure of any project is in your hands. You will learn new tricks by experimentation. Experiment only if you are recording or consciously thinking about what you are doing. You need to be able to reproduce the parameters next time. Keep a logbook, and write up what you did when you finish the job. This way you can do it again next time you get a similar problem.

DRILLED NOZZLE OR NOZZLE INSERTS?
Many of the off-the-shelf bombs you can buy are drilled holes—holes drilled from the outside to the centre. We know about nozzle Cd, or coefficient of discharge. when discussing a nozzle's performance data. This relates to the amount of power required to push the water through the nozzle. We also know something about friction loss, the "thief" that steals the pressure from your pump. So, *if* the 1.0 mm drilled hole is 10 mm long, your Cd is going to drop to about 0.68. That relates roughly to 100 horsepower in and only 68 horsepower out, you are losing 32 horsepower just to get the water through the long hole.

Why the long hole? Three reasons. The first is the wall thickness needed to hold the pressure inside the bomb requires that the drilled hole, at an angle, is drilled through the thick wall.

The second is that the material the bomb is made of, typically 316 stainless (which incidentally has similar hardness and wear resistance to aluminium bronze; for good wear and strength properties use 630 stainless or 420L and have it hardened) has very little abrasion resistance. If the hole was short, the nozzle hole would wear out in a couple of hours, and the bomb would be useless. By making it long, the hole lasts longer.

The third reason is cost. To set up an NC to make bombs with drilled holes is quite simple, and the process requires very little labour. To make the bomb with threaded ports suitable for fitting proper nozzles is a great deal of work and quite costly, without even considering the price of the insert.

A high Cd number requires less pressure to "push" the same number of litres per minute out of the hole. One hundred litres per minute at 1000 bar in a short hole could drop to 100 litres per minute at 750 bar out (available to do work) of the same hole diameter with the hole 1.5 mm longer. Long holes have low Cd. However, long holes have a better jet shape and cohesion; manufacturers try to get the balance of Cd, cohesion, and jet form, which is a bit of a trick to do. So do not ignore nozzles with low Cds; look at what you are doing against how the nozzle performs on the job.

Take a look at a typical step nozzle; the distance from where the taper stops to the outlet of the nozzle is only a couple of millimetres, and the manufacturer claims a Cd of 0.95. Make that parallel section one mm longer, and the Cd drops to about 0.78. Make it another mill longer, and the Cd drops to about 0.60.

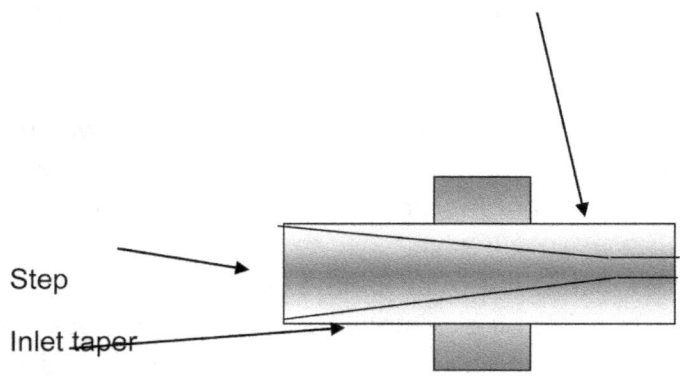

Figure 6.10. A "step nozzle" showing step, inlet taper, and nozzle form

Nozzles and Holders

When choosing a drilled hole size for your bomb, select a diameter about 20 per cent bigger than the nozzle selection chart (say you chose a 1mm dia nozzle drill the hole 1.2mm) indicates for inserts. This is particularly important if the hole is going to be long. All drilled nozzle holes should have a 90 per cent included (45 degree and 45 degree) angle on the outlet point to prevent feathering of the water jet as it leaves the nozzle. Explanation: after the nozzle hole has been drilled, take a drill about twice the hole size, whose tip has been ground to 45 degrees on each side, and just polish the outside lip of the hole. This prevents feathering and keeps the jet form cohesive longer.

Some manufacturers do make excellent bombs with drilled holes. These are properly recessed to shorten the length, and the bomb is made from a harden-able material. However, they are quite expensive and, once worn, have to be thrown away.

Bombs with proper nozzle inserts, normally threaded inserts, with a Cd in the 90s have the holes drilled and threaded, and the correct nozzle selection for your pump is supplied. Once worn, these nozzles are replaceable, and you have a new nozzle holder once again.

The following picture shows big drain cleaners using welded-in, bent, stainless tube, one for each nozzle. This significantly reduced friction losses and gave a much better flow rate than conventional drilled blocks.

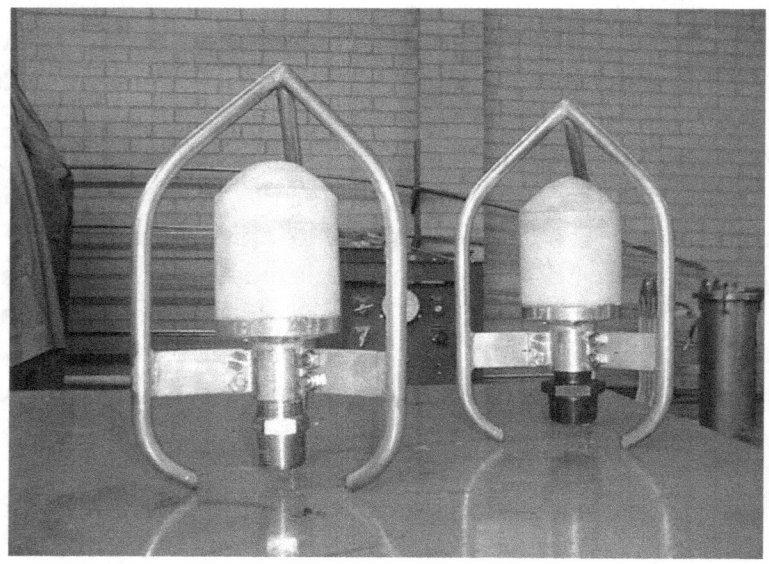

Figure 6.11. Big drain nozzle holders made with bent tube for each nozzle and encapsulated in rubber (manufactured by T-Squared)

Vee, Fan, or Flat Jets?

These nozzles also have a place in drain cleaning. Moving lots of fine sand? Do not fit pin jets. These cut slots or grooves through the product and act like a rake. Fit vees and overlap them. This provides a cone of water acting much like a "scoop" or bucket. Stick to about 15-degree angle on the jets though; greater angles distribute a little water over a large area and really only make things wet. A big aluminium bomb on a ¾" hose with eight or more vee jets will shift a huge amount of product per pass *if* that product is loose and reasonably fine

The best pick of the vee jets, in my opinion, for pressures to 600 bar is the Spray Systems ¼ MEG series. These have a good jet shape and long life, and they are not expensive. Don't bother with the "flow straighteners" if using these on a gun barrel. Flow straighteners are needed only when the supply is a 90-degree turn immediately before the nozzle inlet point and the flow out of the nozzle is turbulent. (Sometimes turbulence is good; it can aid in breaking up muds and clay.) The thread is ¼" NPT, which does not fit well into a ¼" BSPT hole. It will do little harm to drop a ¼" NPT tap through the ¼" BSPT hole, as a ¼" BSPT nozzle or fitting fits quite well into a ¼" NPT hole. If you do not have a nozzle selection chart for vee jets, use the nozzle chart in this book and go up two 0.1 sizes. Select a 2.2 pin jet from the chart? Then fit a 2.6 vee. Vee jets have a low Cd but at lower pressures and high volumes this is not too important.

Figure 6.12. ¼ MEG nozzles, flat, and pin (manufactured by Spray Systems)

The following picture is of step nozzles made by T-Squared. As standard, the nozzle is made in only one outside dimension; it has an 8 mm diameter "step". For some types of tooling, the step needs to be 12 mm, and so a small, cupped bush is provided into which the smaller step is pressed. Different sizes of O-rings are used for either the 8 or 12 mm styles.

The nozzles are made from a special stainless steel (420L) and nitrogen hardened to a hardness of 68 Rockwell. These >96 Cd nozzles will outlast most others and are quite expensive.

Figure 6.13. Step nozzles with bush to cover all needs (manufactured by T-Squared)

Another interesting drain nozzle is the skirt nozzle. This does not have jets but rather one cone fitting over another, creating a restriction that causes the water to exit in a perfect cone. It is not a great cutter but is a great soil mover that works well in pipes with rubber lining that you want to be careful not to damage. Flow rate and pressure is operator adjustable to 1000 bar.

Figure 6.14. Skirt nozzle (design and manufactured by T. Everest)

**Figure 6.15. Vibrating nozzle "Scale Breaker"
(manufactured to order by T-Squared)**

This tool was designed to break up bauxite scale in pipes, operating at a pressure of 250 bar with a 1" hose and 280 litres per minute. The unit weighed 7 kg and was 100 mm in diameter. The teeth are made of tungsten carbide welded into a stainless body. The centre, offset, spinning portion is made from aluminium bronze. It had four nozzles: two straight out to drive the rotation and to break up the scale and two back at 30 degrees to push away the removed product and to act as a speed break.

Designed to be winched up the pipe, it turned out that the tool would propel itself about 15 metres. It is a very impressive tool and seems to melt through hard scale. Spinning at about 200 rpm, the teeth rapidly pound their way through the hard deposits.

It follows that the style of bomb you select depends on your imagination and creativity. Take the time to think about what you want to do, how much time you have, and how far you want to travel. Think of standoff, angle of impact, distance to the pipe wall, and material the pipe is made of. Now go find an engineer you can talk to, and get him to make some to your design.

Chapter Eight

Nozzle Selection and Friction Loss

Selecting the correct nozzle inserts to give you the jet form you need is a skill you *must* have. First you need to know the pump flow and pressure, then divide the number of nozzles into the flow rate, ignoring the pressure at this stage. Assuming a flow rate of 100 litres per minute and five nozzles, you will end up with 20 litres per minute per nozzle. Select the nozzle size from the chart.

If you have forward shooters you will need to do some calculations. Assume you have 1000 bar pressure available. Reaction in kg = 0.224 x litres x √1000 (enter 1000 and hit the square-root button on your calculator). NOTE: 0.0224 is a factor and is used for all calculations of reactive force.

R = 0.0224 x 20 litres x 31.62 = 14.16 kg per nozzle

Say you have two straight forward and three back nozzles. From the above calculation of reaction per nozzle, you have a push of 28.32 kg from the front (2 x 14.16 trying to push the hose back out of the pipe).You have 42.18 kg pushing the hose up the pipe from the back shooters = 13.86 kg of pull after subtracting the front nozzles. Assume a horizontal pipe and ½" hose weight = 0.3 kg metre—13.86 divided by 0.3, and you have 46 metres of travel available to you.

Now assuming that the rear nozzles are at 45 degrees and the front ones are straight, you lose some power to the angle. At 30 degrees subtract 25 per cent; at 45 degrees subtract 35 per cent: 13.86—35% = 9 kg, divided by 0.3 = 30 metres of travel. (See figure in page 95.)

NOTE: The thick, black line running across some nozzle charts indicates a reaction of about 25 kg, to the left of the line hand hold, to the right of the line mount the nozzle on a tool.

	Nozzle Size															
	1.0	1.2	1.4	1.6	1.7	1.8	2.0	2.1	2.2	2.4	2.6	2.8	3.0	3.2	3.4	3.6	3.8	4.0
50	4.2	6.3	8.6	11.1	12.5	13.9	17.2	18.9	21.0	24.5	28.8	33.5	38.3	43.8	49.5	55.4	61.3	68.1
75	5.2	7.7	10.5	13.6	15.3	17.1	21.0	23.2	25.7	30.1	35.3	41.0	46.9	53.6	60.6	67.8	75.0	83.4
100	6.0	8.9	12.1	15.7	17.6	19.7	24.3	26.8	29.7	34.7	40.7	47.3	54.2	61.9	70.0	78.3	86.7	96.4
125	6.7	10.0	13.5	17.5	19.7	22.0	27.2	29.9	33.2	38.8	45.5	52.9	60.5	69.3	78.2	87.5	96.9	108
150	7.4	10.9	14.8	19.2	21.6	24.1	29.8	32.8	36.3	42.5	49.9	58.0	66.3	75.9	85.7	95.9	106	118
175	7.9	11.8	16.0	20.7	23.3	26.1	32.2	35.4	39.2	45.9	53.9	62.6	71.6	81.9	92.6	104	115	127
200	8.5	12.6	17.1	22.2	24.9	27.9	34.4	37.9	41.9	49.1	57.6	67.0	76.6	87.6	98.9	111	123	136
225	9.0	13.3	18.1	23.5	26.4	29.6	36.5	40.2	44.5	52.1	61.1	71.0	81.2	92.9	105	117	130	145
250	9.5	14.1	19.1	24.8	27.8	31.2	38.4	42.3	46.9	54.9	64.4	74.9	85.6	97.9	111	124	137	152
275	10.0	14.8	20.1	26.0	29.2	32.7	40.3	44.4	49.2	57.6	67.5	78.5	89.8	103	116	130	144	160
300	10.4	15.4	20.9	27.1	30.5	34.1	42.1	46.4	51.4	60.1	70.5	82.0	93.8	107	121	136	150	167
325	10.8	16.0	21.8	28.2	31.7	35.5	43.8	48.3	53.5	62.6	73.4	85.4	97.6	112	126	141	156	174
350	11.2	16.7	22.6	29.3	32.9	36.9	45.5	50.1	55.5	64.9	76.2	88.6	101	116	131	146	162	180
375	11.6	17.2	23.4	30.3	34.1	38.2	47.1	51.9	57.4	67.2	78.8	91.7	105	120	135	152	168	187
400	12.0	17.8	24.2	31.3	35.2	39.4	48.6	53.6	59.3	69.4	81.4	94.7	108	124	140	157	173	193
425	12.4	18.3	24.9	32.3	36.3	40.6	50.1	55.2	61.1	71.6	83.9	97.6	112	128	144	161	179	199
450	12.7	18.9	25.7	33.2	37.4	41.8	51.6	56.8	62.9	73.6	86.4	100	115	131	148	166	184	204
475	13.1	19.4	26.4	34.1	38.4	42.9	53.0	58.4	64.6	75.7	88.7	103	118	135	152	171	189	210
500	13.4	19.9	27.0	35.0	39.4	44.1	54.3	59.9	66.3	77.6	91.0	106	121	139	156	175	194	215
550	14.1	20.9	28.4	36.7	41.3	46.2	57.0	62.8	69.6	81.4	95.5	111	127	145	164	184	203	226
600	14.7	21.8	29.6	38.4	43.1	48.3	59.5	65.6	72.7	85.0	99.7	116	133	152	171	192	212	236
650	15.3	22.7	30.8	39.9	44.9	50.2	62.0	68.3	75.6	88.5	104	121	138	158	178	200	221	246
700	15.9	23.5	32.0	41.4	46.6	52.1	64.3	70.9	78.5	91.8	108	125	143	164	185	207	229	255
750	16.5	24.4	33.1	42.9	48.2	54.0	66.6	73.3	81.2	95.1	112	130	148	170	192	214	237	264
800	17.0	25.2	34.2	44.3	49.8	55.7	68.7	75.7	83.9	98.2	115	134	153	175	198	221	245	273
850	17.5	25.9	35.3	45.7	51.3	57.5	70.9	78.1	86.5	101	119	138	158	181	204	228	253	281
900	18.0	26.7	36.3	47.0	52.8	59.1	72.9	80.3	89.0	104	122	142	162	186	210	235	260	289
950	18.5	27.4	37.3	48.3	54.3	60.7	74.9	82.5	91.4	107	125	146	167	191	216	241	267	297
1000	19.0	28.1	38.2	49.5	55.7	62.3	76.9	84.7	93.8	110	129	150	171	196	221	248	274	305
1050	19.5	28.8	39.2	50.8	57.1	63.9	78.8	86.8	96.1	112	132	153	175	201	227	254	281	312
1100	19.9	29.5	40.1	51.9	58.4	65.4	80.6	88.8	98.4	115	135	157	180	205	232	260	287	320
1150	20.4	30.2	41.0	53.1	59.7	66.8	82.4	90.8	101	118	138	161	184	210	237	265	294	327
1200	20.8	30.8	41.9	54.3	61.0	68.3	84.2	92.8	103	120	141	164	188	215	242	271	300	334
1250	21.2	31.5	42.8	55.4	62.3	69.7	85.9	94.7	105	123	144	167	191	219	247	277	306	341
1300	21.7	32.1	43.6	56.5	63.5	71.0	87.6	96.6	107	125	147	171	195	223	252	282	312	347
1350	22.1	32.7	44.4	57.6	64.7	72.4	89.3	98.4	109	128	150	174	199	228	257	288	318	354
1400	22.5	33.3	45.2	58.6	65.9	73.7	90.9	100	111	130	152	177	203	232	262	293	324	361
1450	22.9	33.9	46.0	59.6	67.1	75.0	92.5	102	113	132	155	180	206	236	266	298	330	367
1500	23.3	34.5	46.8	60.7	68.2	76.3	94.1	104	115	134	158	183	210	240	271	303	336	373
1550	23.7	35.0	47.6	61.7	69.3	77.6	95.7	105	117	137	160	186	213	244	275	308	341	379
1600	24.0	35.6	48.4	62.7	70.4	78.8	97.2	107	119	139	163	189	217	248	280	313	347	385
1650	24.4	36.2	49.1	63.6	71.5	80.0	98.7	109	120	141	165	192	220	252	284	318	352	391
1700	24.8	36.7	49.9	64.6	72.6	81.2	100	110	122	143	168	195	223	255	288	323	357	397
1750	25.1	37.2	50.6	65.5	73.7	82.4	102	112	124	145	170	198	227	259	293	327	363	403
1800	25.5	37.8	51.3	66.5	74.7	83.6	103	114	126	147	173	201	230	263	297	332	368	409
1850	25.8	38.3	52.0	67.4	75.7	84.8	105	115	128	149	175	204	233	266	301	337	373	414
1900	26.2	38.8	52.7	68.3	76.8	85.9	106	117	129	151	177	206	236	270	305	341	378	420
1950	26.5	39.3	53.4	69.2	77.8	87.0	107	118	131	153	180	209	239	274	309	346	383	425
2000	26.9	39.8	54.1	70.0	78.8	88.1	109	120	133	155	182	212	242	277	313	350	388	431

Figure 6.16. Nozzle selection chart

You may need to consider decreasing the size of the forward jets and increasing the size of the back jets if you have long runs. The forward thrust from the back jets must always be at least 30 per cent more than the back thrust of the forward jets. If you have two forward and two back jets, one of the back jet's reactive force *must* be equal to or greater than the sum of the two front jets, in case you get a blockage.

SELECTING THE NOZZLE

Looking at the nozzle chart, you find the system pressure marked off in bars down the left side (1 bar = 14.5 psi). Run a ruler along the horizontal column until you arrive at your required flow rate per nozzle. Run your ruler up the page until you find the nozzle diameter line. The figure is the diameter of the nozzle you need in millimetres.

The chart is for nozzles with a Cd (coefficient of discharge) of between 0.90 and 0.96. In short, this indicates the power coming out of the nozzle, typically 100 kw in, 96 kw out. There are many styles of nozzle and each has a different Cd. You need to know the Cd of the nozzle you are using. Say, for argument's sake, your nozzles have a Cd of 80 per cent. You might choose the next nozzle size up. If the chart shows a 1.2, select a 1.3 to adjust for the Cd difference.

HOSE FRICTION LOSS IN NOZZLE SELECTION

Another thing to remember is friction loss through your hose. Absolutely spot-on nozzle selection requires you to subtract the friction loss *before* you select the nozzle size. *The pump pressure gauge does not indicate the pressure at the nozzle.* If you really want to know what pressure you have at the gun end or nozzle holder, take out the nozzles, run your unit up to its normal working RPM, and read the gauge. The pressure shown on the gauge is what it costs to get the water to the nozzle.

My example above said 100 litres and went on to work it through ½" hose. The attached chart will help to do this.

Nozzle Selection and Friction Loss

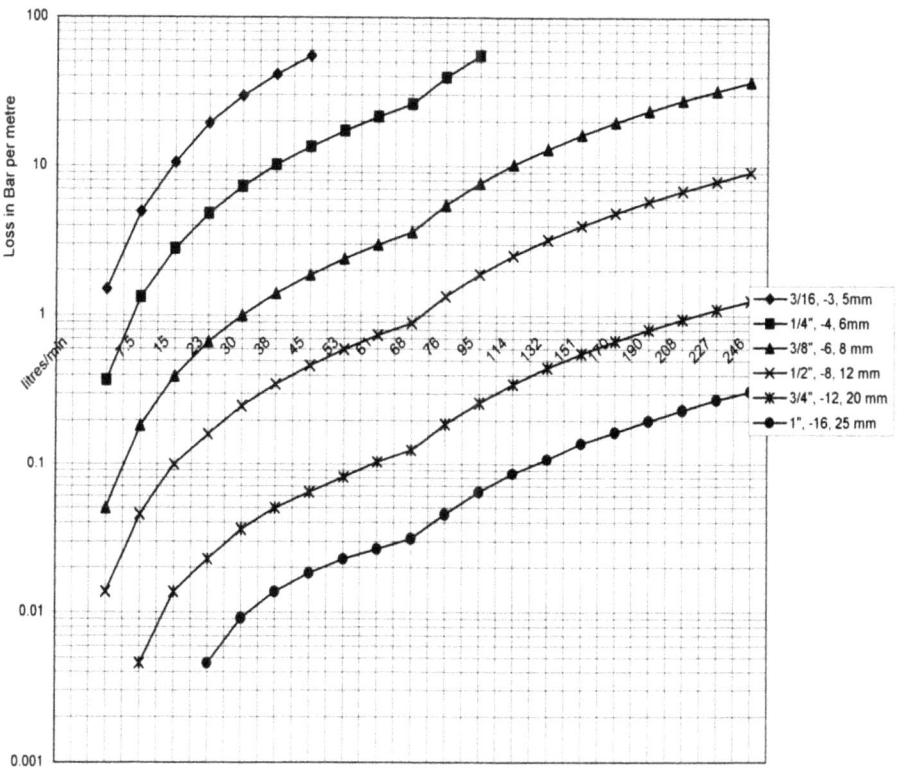

Figure 6.17. Hose friction loss chart

Across the centre of the chart is a row of numbers: litres/min. Pick a point between 95 and 114 where 100 should be, place a ruler vertically up the page and read off the ½" hose line. Your loss is 2 bar per metre of hose. A 20 metre length will have 40 bar lost to friction on a new hose. If it is getting a bit old, add 20 per cent so the loss now equals 48 bar. Subtract that from 1000, and you have 952 bar available at the nozzle. If you have two hoses out, the loss will be 100 bar, and so on.

If possible, try to select a hose to keep your friction losses below the centre line. Less than 1 bar/metre is very good if you can get it.

Friction loss is a killer. Let's look at those areas where friction can eat up your available pressure.

Say you have 100 meters of hose on the reel:

½" hose at 60 litres, loss = 75 bar (1087 psi)

- ½" hose at 100 litres, loss = 200 bar (2900 psi)
- ½" hose at 80 litres, loss = 150 bar (2175 psi)
- ¾" Hose at 120 litres, loss = 40 bar (580 psi)
- ¾" hose at 150 litres, loss = 55 bar (795 psi)
- ¾" hose at 180 litres, loss = 70 bar (1015 psi)
- 1" hose at 200 litres, loss = 20 bar (28,5 psi)
- 1" hose at 250 litres, loss = 30 bar (43.5 psi)
- 1" hose at 300 litres, loss = 50 bar (72.5 psi)

The above are just some indications of your friction losses with *new* hose. Be aware of this, and remember that the pressure on the gauge attached to the pump tells you only the pressure at the head and not at the nozzle.

An interesting test is to remove the nozzle and run the pump up to operating speed. The pressure registering on the gauge is the "cost" in friction to get the water to the bomb or gun. You may see figures as shown above; start your calculations, and try to reduce the losses.

THINGS TO LOOK AT TO REDUCE FRICTION LOSS

The pressure regulator can cost as much as 12 per cent of your available pressure. This unit *must* be located on the other side

Nozzle Selection and Friction Loss

of the pump. Your hose water does not need to pass through it; only the water going back to the tank or to the dump should pass through the regulator.

A 90-degree elbow will cost you between 4 and 7 per cent of your pressure. There should not be one anywhere in the circuit. If you have one, get it out and replace it with a looooong bend.

A hose swivel on the hose reel will cost between 5 and 7 per cent, unless it is a really big one with lots of space for the water to move in. If you are using ¾" hose, you should have a 1 ¼" swivel, not a ¾" one.

Hose reeled onto the hose reel that was not pressurised when it was wound on will cost you as much as 35 per cent of your available pressure. Watch this one most carefully. As pressure comes on to the hose, the hose shrinks or shortens. The outside diameter of your hose gets bigger, and the length is reduced. When you pressurise the reel, this movement can flatten the bottom layer of hose and, if not coiled on properly, the pressurised hose can crush the drum in seconds.

Attach your garden hose to the tap, turn the tap on, and squeeze the end of the hose. What happens? You get a longer stream of water at greater speed, yes? What happened inside the hose? The pressure increased because the water could not get out as easily. Friction loss created by the deformed hose caused the pressure to back up. The hose is still the same area, but it is no longer round; there is a big difference in flow rate through a flattened hose and a round one.

When fitting a new hose to a hose reel, plug the end and set the regulator at pump pressure and then wind the hose on. When you let the pressure off, the hose on the reel will look a little untidy, but as soon as you pressure up, it will all straighten out again.

If you have a kink in your hose and this is trapped in amongst the hose wraps, the losses could be massive, say 90 per cent. Uncoil the hose all the way, pressurise it, and wind it back on with care.

Flat-ended hose joiners and fittings can cost you as much as 2 per cent each and be destroyed by cavitation. Taper the inlet and outlet edges. Do not let the water see flat faces.

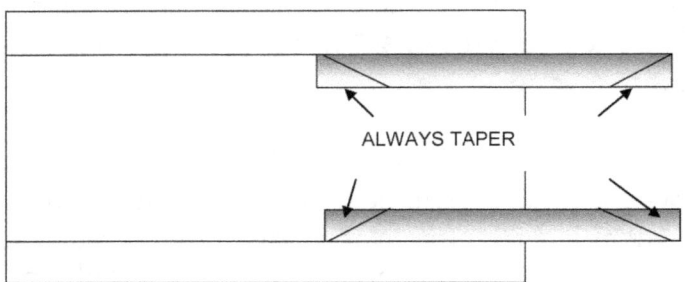

Figure 6.18. Where the water flow meets a flat face please taper the step.

Old hose can treble the new hose-friction loss factors. The friction loss chart indicates the loss to new hose. It is a good idea to reverse the hose on the reel periodically. Small dags (three-cornered tears) in the hose wall that lift up into the water stream causing a restriction are knocked down flat when the flow is in the other direction.

The pipe-cleaning bomb will eat up about 8 per cent of your pressure. The water has to make a dramatic turnaround to get the stream pointing back into the nozzles. Some pipe bombs, made with coiled stainless pipe cast into rubber or aluminium, have very little friction, but these are large diameter bombs, as space is taken up by the bent tubes. If you are working in big pipes, this style of bomb is a valuable investment.

Nozzle Selection and Friction Loss

Look at your plumbing and reel supply circuit. Try to remove all restrictions between the pump and the swivel. Have a careful look at the point where the hose on the reel is attached to the metal fitting from the swivel; this should have no sharp bends or kinks in it.

If you have a wet-dump foot valve in the line, set it off to one side, "tee" it off the main line, and let the water travel unrestricted from the pump to the bomb.

The idler roller where the hose runs off the reel over the side of the truck should be curved to fit the hose diameter. A flat roller will flatten the hose and cause a restriction.

Each little thing you do will give you just a little bit more "grunt" to help get the job done quicker and beat the competition.

A good indication of friction somewhere in the system is heat. If your water is heating up, it has cost you energy to generate that heat. When I first started water jet work, I thought that, as water will not compress, what was shown on the gauge was the system pressure all the way to the nozzle. It is, *if* the end of the hose is blocked off and the water is not flowing. However, once it starts to flow, it is pushing past all manner of obstructions in its effort to get out. Each one of these obstructions, not unlike sandpaper, is rubbing on the water, trying to make it slow down. To get past the obstruction, energy is required. This energy becomes heat. That heat is pressure lost.

If you imagine the energy required by your domestic hot-water heater to heat a bath full of water, you will get some idea of the energy needed to heat up a continuous flow of water through a set of plumbing, triggers, foot valves, and old hose. This is not free energy; it has a cost. That cost is jet velocity and horsepower and ultimately fuel for your engine and working time. Friction can prevent you doing your job; eliminate it wherever you possibly can. One extra hose length or 10 more litres per minute through a small hose can make the difference between the success and the failure of your job.

For the record, water *does* compress. At 2000 bar, or 30,000 psi, it compresses about 15 to 17 per cent of its volume.

Chapter Nine

Designing a Drain or Pipe-Cleaning Holder

For this exercise, we will assume that we have a pump capacity of 345 bar (5,000 psi) with a flow rate of 250 litres per minute. Please remember that the pressure gauge on your pump tells you only the pressure at the pump and bears little relationship to the pressure available at the nozzle.

Check friction losses through your hose, and calculate the pressure available *at the hose end*.

1. Decide on the number of nozzles and angle.

2. Decide on the nozzle bore diameter.

3. Decide the diameter of your bomb.

4. Decide on the jet angle.

We have a 60 metre drain pipe, a quarter filled with road gravel bound up in sand and soil—a reasonably tough job. The pipe is 600 mm in diameter and made of concrete. It is in good order with no concrete degradation.

We have decided that we will have one forward jet at 5 degrees off straight and six back-facing jets at 45 degrees.

The bomb will be 60 mm in diameter and 80 mm long and made of engineering-quality aluminium. That means that the nozzles we use will need to be threaded inserts, as aluminium will wear out rapidly if the holes are simply drilled. The nozzles will need to be drilled and tapped into the side of the bomb. We cannot drill into the back face, as we will cut into the 1" female hole for the hose.

Ideally the holes drilled for the nozzles should be ¼" NPT for Spray Systems's ¼ MEG pin or vee jets, if vee jets get 15 or 25 degree angle—but no larger. Their part number is normally four figures with the first two being the angle 15XX and the second two being the bore.

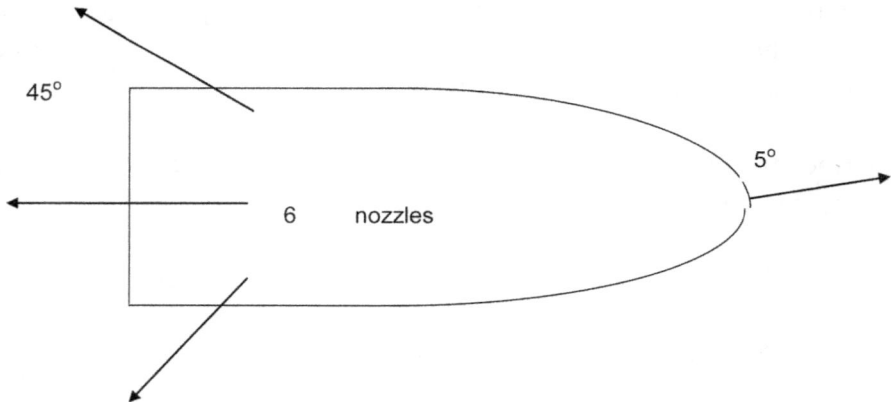

Figure 7.1. Typical bomb configuration

Now, 250 litres divided by 7 equals 36 litres per minute per nozzle. We will assume that we will have all the nozzles the same size at this stage. Do not divide the pressure; divide *only* the flow.

Go to the nozzle selection chart and look for 345 bar—the nearest is 350 bar. Looking along that line, we see that a 1.8

Designing a Drain or Pipe-Cleaning Holder

nozzle will give us 36.9 litres per minute at that pressure. We will accept 36.9 as the correct flow rate for this project. This will take up the little bit of friction created by the nozzle head or bomb turning the water back along the hose line.

The reactive force from each nozzle is $0.0227 \times 36 \times \sqrt{345} = 15.2$ kg.

Our back pushers are angled at 45 degrees, so we need to subtract the loss to the angle. (See figure in page 95.) This will be 25 per cent off the available force for an angle of 45 degrees. So, we now have 11.38 kg per back-facing nozzle. Multiplied by six nozzles, we have 68 kg available to pull the hose. We must subtract the thrust pushing back from the front nozzle, so 68.3 minus 15.2 = 53.1 kg of total pull available.

Assume we are using 1" rubber drain hose weighing 1.5 kg per metre in air. Lying flat in a clean pipe, it will require about 0.25 kg per metre to move it: 0.25 kg divided into 53 kg = 212 metres. We have plenty of pull to travel through our 60 metre long pipe above.

However we boobed. We did not allow for friction loss through the hose. This is a common mistake.

We need to look at the friction losses through the 1" hose, as this could blow all our calculations. Go to the friction chart and look up 250 litres. It is just off the chart, but if we mentally extend the line to about where 250 litres would be, we can see that the friction loss is going to be about 0.4 bar per meter. If we had 100 metres of hose on the reel, we have a problem. It has blown our maths. We have only 305 bar available at the nozzle.

Back to the drawing board. The nozzle chart at 300 bar now indicates 2.0 mm nozzles giving us 42.1 litres per minute each. Multiplied by 7, that is 294 litres—44 litres too much. With 1.8 mm nozzles we would still be about right—about 12 litres too small in total; 1.9 mm would be spot on.

Has anything happened to our reaction force available to pull the hose?

- 0.0227 x 34 x √245 = 14.1 kg per nozzle x 6 = 84 kg

- Less 25 per cent for the angle = 63 kg

- Less the forward jet at 14.1 kg = 48.9 kg available

- Weight of our hose in horizontal pipe is 0.25 kg/m = 195.6 m

We got away with it this time. We still have plenty of pull fortunately, and we can build our bomb.

I deliberately "forgot" the friction-loss calculation just to show how easy it is not to allow for it and how it can sometimes screw things up. If we had been using ¾" hose, the loss would have been about 1.5 bar per metre—150 bar lost. All we would have had available for our bomb would have been 195 bar. That's scary stuff, but often overlooked.

Remember, friction losses on the chart are for new hose in a straight line. Losses build up as hoses age, fittings are inserted, and hoses are coiled.

Whenever you are selecting nozzles *always* subtract the friction loss from the total pump pressure first. Again, all the pump pressure gauge is telling you is the pressure available at the pump head and *not* the pressure at the hose end.

As an added exercise, let's look at the amount of spoil we are going to get back. First of all, we need to know the height the spoil is from the bottom of the 60 metre, 600 mm pipe. We know it is a quarter full—that is, 150 mm. A measurement across the top of the spoil would be about 450 mm. The volume therefore is 0.075m x 0.45m = 0.03 m^2 x 60 = 2.025 cubic metres, or about 5 tonnes of spoil out of the pipe (width across

Designing a Drain or Pipe-Cleaning Holder

the top multiplied by half the depth). This does not include the weight or mass of the water used.

What metal do we use to make the bomb? You have several choices, but here are some options:

1. Aluminium for pressures up to 300 bar. (Use engineering-grade machining aluminium, not melted down coke cans.) Be aware that some acids eat aluminium, so check whether it is going to work. These will need threaded nozzle inserts, so the smallest diameter should be 60 mm.

2. Aluminium bronze, a mixture of aluminium and bronze (called alibronze), has excellent wear properties and will hold pressures to about 600 bar. It is as strong as 316 stainless but wears better. The drilled jet holes wear quite quickly at higher pressures, so drill and tap as for aluminium. For low pressures, such as the single cylinder engine units, the holes can be drilled and the bomb will last a long time.

3. 316 stainless steel is, in fact, less long-lasting than alibronze, which is harder and has a higher yield strength (pressure at which it will burst). So 316 is "overrated" as everybody's idea of the optimum material. It is optimum if you are buying yacht fittings, but as a high-pressure carrier, it is not really acceptable. When using 316, ensure that the minimum wall thickness excluding the depth of the thread is at least 6 mm. Then restrict it to pressures to 630 bar, or 10,000 psi.

4. 630 stainless is quite hard and the best material for higher pressure nozzles if you are in a hurry. It can be drilled for nozzles but will wear over weeks of work. Pressures to 1400 bar are acceptable with 630, but keep the wall thickness at least 6 mm, as for 316. There are two types of stainless bar stock in most grades; always insist on "machining quality". Some are designed primarily for shafting and are very

difficult to work with (broken drills and taps are common with these). As a rule of thumb, all stainless with a 3 in the number is not hardenable after machining.

5. 420 and 420L stainless is the best to machine while unhardened. Once hardened, drills will burn before they will mark it. It hardens up to tool steel, close to tungsten carbide, and will last for years. The material is cheap to harden; in Perth it was recently 25-dollar per kg to harden. Hardeners do not charge by the item; they charge by weight of the lot. If aesthetics are important to you, it can be vacuum-oven hardened so that the finished product remains silvery and shiny. Ordinary hardening darkens the metal. Most water jet nozzles are made of 420L and hardened in a vacuum oven.

6. Titanium, also excellent, is essential in nickel process pipe cleaning, as it will resist extreme acids. Titanium is almost as light as aluminium, which helps in pipe travel. It is not as expensive as we are led to believe, but the biggest problem is availability. If there is an aircraft breaker near you, see if you can buy scrap wheel struts; these are titanium. Check some acid process plants for big bolts or studs lying about. Ask permission though, as people get quite excited about people picking up their scrap. Remember to use lots of lathe coolant when machining titanium; the shavings or swarf can burst into flame and are difficult to extinguish.

7. Some hard plastics are now available. I have made several big drain bombs for pressures up to 300 bar out of it, and the threads hold without leaking. Also, the bomb does not hurt the pipe when tossed into a manhole clumsily. There are many kinds of plastic, and each type is hidden by a trade name, so talk to a local supplier. Use a really sharp and new tap for the threads, as damage to the body will occur if the tap does not slide in easily. Prices compare favourably with metal stock.

Designing a Drain or Pipe-Cleaning Holder

Bomb - hose connecting threads

Most hose ends are parallel threads—that is, the width of the thread is the same along the length. These do not screw into tapered threads properly; if you get two turns, you're lucky. We need a minimum screwed-in depth of half the diameter, preferably more, to get a good seal. So we make the thread in the bomb parallel too. We machine the bottom of the threaded section flat and put in a copper washer for the hose tail to tighten up onto.

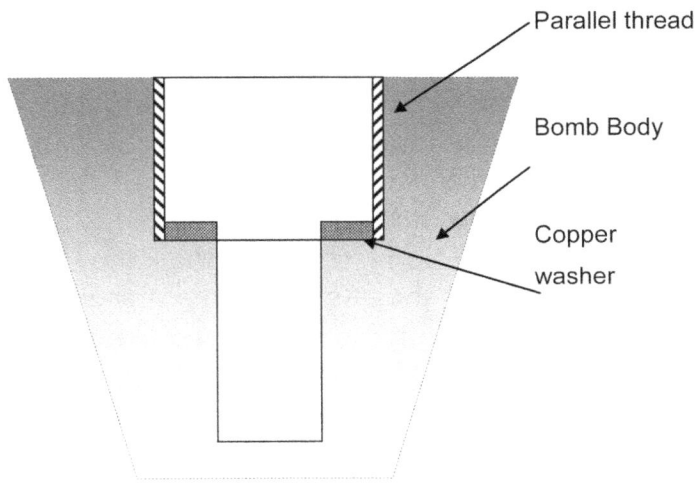

Figure 7.2. Bomb-to-hose thread connection suggestion

Always assemble using anti-seize on stainless to stainless if you want to get it off eventually. Do not use thread tape on any high-pressure application. Instead, use either Loctite 567 or 667 thread sealant. It's great stuff that seals well, and you can get the thing apart later.

Leaks at the fitting are most unsightly and unprofessional. Do it up right the first time. Stop as soon as you see a leak; if you don't, you will wear a groove in the bomb or the fitting, rendering them both useless. Never tighten a fitting under pressure, or it will burst and injure you. Stop the pump, depressurize the system, and then fix it. *Please* do not use pipe wrenches; they wreck fittings, and the dags will cut you.

Remember at all times that the water jet industry is allowed to operate with a safety factor (working over burst) of 2.5 to 1. Pipe wrenches cut grooves, reducing the safety factor to a dangerous level. The norm for hydraulic and gas fittings is at least 4 to 1. We could not carry the weight of a gun, lance, or hose rated at 4 to 1, so we are allowed to reduce the ratio. But we must take care of the equipment and *inspect* it regularly for damage. Any suspect tube, fitting, hose, or component must be removed from service. Ninety-nine per cent of all our tooling is engraved with its maximum working pressure. Check this against your bursting disc rating.

The shape of the (bomb) nozzle holder
Bombs do not have to be bomb shaped. I think they look good with a tapered point, but those shaped like a shovel, ball, or beer can, and round ones on a shaft shaped like a 4 lb. hammer will also do the job. Use your imagination and be creative. I recently saw some drain-cleaner bombs made out of old breathing apparatus bailout bottles about the size of a big coke bottle. They were light, high tensile, and a nice shape for travel. Make a bomb you consider best suited to the job you are going to do. Above all, get the jets pointing where they will do the most work, and do not waste water, energy, or fuel by inserting jets that will simply blast into space.

Remember also standoff, matchbox, and half brick, and angle of attack. For drilled and tapped bombs, normally ¼" BSP or NPT, you can buy male/female fittings. The male screws into the bomb, and the nozzle screws into the female. You have just extended you nozzles 22 mm closer to the pipe wall. If you are going to skid mount, you can add as many fittings as you need to get the jet from a small bomb close to the wall where the jet is needed.

Designing a Drain or Pipe-Cleaning Holder

If you do fit forward-facing jets to your bomb to get through a blockage, try to tilt them slightly by drilling 3 or 5 degrees off line, which is enough. When you reach the blockage, rotate your hose and blast a hole that is big enough to get the bomb through, so the back jets can get at the deposit and blow it away. With one forward-facing jet dead straight, you can get the jet to blast through; but the bomb cannot follow, as the jetted hole is too small.

This offset will tend to turn the travelling bomb slightly side on to the pipe. No problem; some of the back jets are now at a really good angle to do work. Want to go up a tee or round an elbow? The offset front nozzle will help you do that. Remember, the hose will not twist; line the offset up with the hose marking, and use this to steer your nozzle.

VERY IMPORTANT NOTE: Please be aware of the potential for the bomb to turn around in the pipe and come back out and kill you. It will come out like a rocket. Make sure the rigid section of the hose tail and bomb is longer than the diameter of the pipe. If not, add a bit of tube between the two. Use back-out preventers if at all possible.

Chapter Ten

Fittings Seals and Joints

High pressure water jet threads and fittings are extremely varied, and leaking fittings cost the plant owner a lot of money in lost time, replacement, and potential injury. As a result of this, they deserve a chapter of their own.

In the main, the industry uses threads for joints up to about 1400 bar; after that, various methods of face-to-face seals are used. The faces are mated or pulled together using threads, but the thread is not expected to hold pressure, only tension.

A simple calculation will give some indication of the forces acting on a 12 mm or ½" fitting under 1000 bar pressure. It equates to about 3300 kg of load, about the weight of a small truck, hanging onto a ½" thread, often screwed in only about four turns. Spooky stuff when you think of the times you have sat on a fitting, pulled a hose up between your legs, or wrapped it around your kidneys.

We talk a great deal about hose care, but the poor old fitting is badly neglected and has tremendous potential to injure the operator. First we will set about identifying the threads we use, and then we will discuss joining them together.

Threads
There are four dominant threads:

BSP, or British Standard Pipe Thread
This thread comes in P for parallel or T for tapered. The thread is similar to the American NPT (National Pipe Taper) in appearance, but only the ½" will actually fit and only at reduced pressures. Both NPT and BSPT or tapered thread sealing is accomplished by thread distortion; in other words, the tapered thread acts somewhat like a wedge and forces the female thread to distort and hence makes a seal. In all cases, a thread sealant must be used. If stainless, an *approved* stainless anti-seize paste *must* be used. Tapered threads do not seal well on second use or reuse.

BSPP or Parallel Thread
With this thread, sealing is accomplished on a seat ahead of the fitting. The forward end of the male fitting may be tapered (hose tail 30 degrees), flat faced with an O-ring groove cut into it, or it seated onto a copper washer. The female is often a hose tail known as a female swivel, which sits inside the flare of the male. Another BSPP seal is a Doughty seal, fitted outside the fitting under the nut. As the male nut reaches the female face, it traps a seal between the two faces and crushes it. It is a great seal for low pressures but not acceptable under the pressures we use, as it will extrude out sideways and needs to be captured. This seal is available in steel or stainless steel.

If the Doughty is used in the bottom of a female hole in place of a copper washer, it can be used for higher pressures and seals quite well.

Metallic body (steel or stainless steel) with a buna rubber seal moulded in the centre.

Fittings Seals and Joints

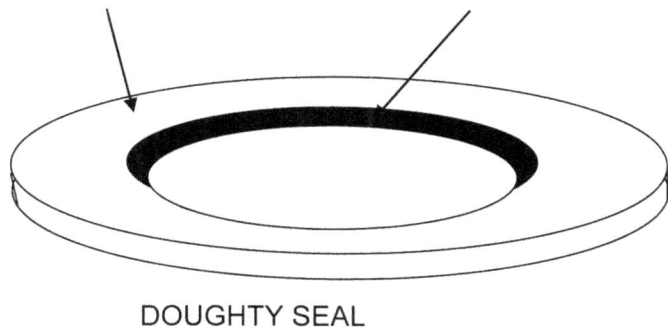

DOUGHTY SEAL

Figure 9.1. Example or a Doughty seal used for sealing fittings

Being able to identify a thread accurately can save you a great deal of time and messing about. I once had an operator ask me for a 36 mm fitting. I had no idea what he was talking about. It turned out that he had taken a crescent wrench and fitted it to the hex section, measured the gap with a ruler—hence 36 mm. He had noticed that the two threads were different on either side. It turned out he wanted an M24 x 1.5 hose tail male to ½" BSPP male adaptor.

As previously stated, there are four dominant threads (actually there are in excess of 300 styles of thread). The following charts will help you to identify some of them.

If you are really keen, pal up to your Aeroquip salesperson. They have a great little reference book and calliper as a sales aid.

BRITISH STANDARD PIPE THREADS—BSP (T for tapered, or P for parallel)

Inch Size	Dash Size	Threads per inch Nominal thread	MALE THREAD OD INCHES Fraction	Decimal	FEMALE THREAD ID INCHES Fraction	Decimal
1/8	-2	1/8 -28	3/8	.38	11/32	.35
1/4	-4	1/4 -19	21/32	.52	15/32	.47
3/8	-6	3/8 -19	21/32	.65	19/32	.60
1/2	-8	1/2 -14	13/16	.82	3/4	.75
5/8	-10	5/8 -14	7/8	.88	13/16	.80
3/4	-12	3/4 -14	1 1/32	1.04	31/32	.97
1	-16	1-11	15/16	1.30	1 7/32	1.22
1 1/4	-20	1 1/4 -11	1 21/32	1.65	1 9/16	1.56
1 1/2	-24	1 1/2 -11	1 /78	1.88	1 25/32	1.79
2	-32	2 -11	2 11/32	2.35	2 1/4	2.26

Figure 9.2. BSP thread table

The "Inch Size" section is obvious. The "Dash Size" is the code used by all hose and fittings manufacturers and is derived by dividing the hose or fitting into sixteenths of an inch, which provides the dash number. Dash or –2 is 2 x 1/16 = 1/8", -4 x 1/16 = 1/4", and so on. The Nominal Thread Size indicates its identification size and the -28 to -11 indicate the number of threads to the inch. If you had a BSP thread gauge and a vernier, you could measure the outside diameter of the thread (about halfway along, if tapered) and then select the thread gauge that fit snugly, excluding all light. You would read off the OD and number of threads per inch and accurately identify the fitting.

Ideally, BSPT should not be used above 320 bar. The British standards do not show it going higher than about 200 bar. However, this is a safety factor in excess of 4:1, and we allow ourselves a factor of 2.5:1, which effectively means that we should not exceed 320 bar operating pressure, or about 5,000 psi. The industry uses BSPP (parallel) for pressures up to 1000 bar, but seals it down onto a copper washer at the face. This ensures that the thread does is not exposed to the pressure; all the thread does is squeeze the copper washer.

The strange thing is that NPT (National Pipe Tapered, United States), which has similar dimensions, is rated for 413 bar with a safety factor of 4:1. In dimensions, the fittings are similar; the material of manufacture is identical; and the threads are close, particularly on the ½″. It really is a factor of calculation, accepted application, and in some cases sheer bloody mindedness on the part of some boffins who were asked to set a standard—one in the United States and the other in Great Britain—and one was too chicken to stick his neck out. That's my opinion and should not be taken as gospel.

A problem with both tapered threads is that as you tighten the male into the female, the female stretches. Then you add, pressure and the female stretches a bit more and lets the water out. The female should be really chunky for HP duty, and wall thicknesses in the range of 6 mm should be searched for. Much less than that and the fitting expands under pressure and the thread ceases to grip or seal.

A term you have possibly heard used when referring to pipe is "half-inch nominal bore". It is half-inch in *name* only (from Latin *nomen*). A half-inch pipe is not exactly half an inch in internal diameter. There are as many as six different "holes through the middle" diameters, all of which are termed ½″ nominal bore pipe. The reason for this is that the outside thread of the pipe (13/16″) must be the same size, while the wall thicknesses change for different pipe pressures. If the outside diameter changed, we would need a heap of different dies and taps to cut threads on the outside of a half-inch pipe.

Pipe is identified as internal nominal bore. *Tube,* on the other hand, is identified as outside diameter. Confused? Copper pipe is tube and is measured on the outside. All our HP lances are tube and are identified by measuring the outside diameter *not* the hole through the middle. The proper description for a lance would be 9/16" tube for 60,000 psi duty. If the lance is made in United States for lower pressures, it could have an NPT left-hand thread on the end. In Europe, it would have either BSPP or metric left-hand threads. Both would have male coned ends for that pressure. (See "Autoclave Fittings" below.)

NATIONAL PIPE TAPERED FUEL—NPTF (NPT)

Inch Size	Dash Size	Nominal Thread Size	Male OD		Thread Inch		Female..		Thread Inch	
			Fraction		Decimal		Fraction		Decimal	
⅛	02	⅛ -27	13/32		.41		⅜		.38	
¼	04	¼ -18	17/32		.54		½		.49	
⅜	06	⅜ -18	11/16		.68		⅝		.63	
½	08	½ -14	27/32		.84		25/32		.77	
¾	12	¾ -14	1 1/16		1.05		1		.98	
1	16	1 -11½	1 5/16		1.32		1¼		1.24	
1¼	20	1¼ -11½	1 21/32		1.66		1 19/32		1.58	
1½	24	1½ -11½	1 29/32		1.90		1 13/16		1.82	
2	32	2 -11½	2 ⅜		2.38		2 5/16		2.30	

Figure 9.3. NPT thread table

US National Pipe Tapered Thread

The measurement and discussion for this thread is similar to that for BSP above. You will note that NPT is available only as a tapered thread. British Standard Pipe is available as either tapered "T" or parallel "P".

A parallel thread set down onto an O-ring or copper washer can carry much greater pressures than can a tapered thread. The reason is obvious if you think about it. With the O-ring or copper washer fitting, the pressure is contained inside the bore of the fitting. Water does not touch the threads; all the threads do is hold the fitting and O-ring firmly between two faces. The tapered thread needs to be cranked down hard to deform the threads and create the seal. The parallel thread just needs to be nipped up to compress the O-ring or seal ring.

Metric Threads (M)

Metric tube and pipe fittings get a little complicated, because there are high-pressure and low-pressure rated fittings in the same size range. For the sake of this discussion, we will discuss only the high-pressure fittings normally prefixed s.Rh. In sizes below about 25 mm, the threads are normally "metric fine". For instance, the M24 fitting would be described as "by 1.5" (s.Rh M24 x 1.5 is the most common thread used in the water jetting industry).

It gets a little confusing here. In BSP and NPT, the threads are described in number of threads per inch along the length of the fitting, in metric the measurement is from the top of one thread to the top of the next. This is hard to measure accurately with a vernier micrometer; it's best use a thread gauge.

The hose joining fitting is normally "Tapered Nose 24° included angle" or "Globeseal". Included angle means that the fitting is tapered 12 degrees on each side and 2 x 12 degrees = 24 degrees. Both angles are included in the number.

In the water jet industry, the tapered Globeseal face (inside the female swivel nut) is fitted with an O-ring that fits inside the 24-degree male fitting.

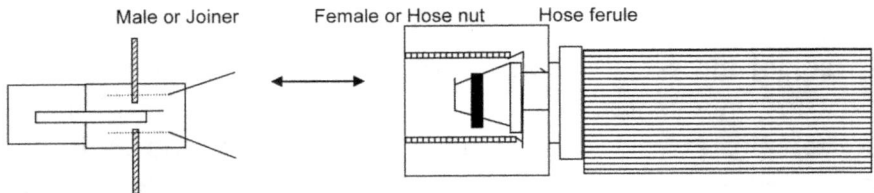

Figure 9.4. Sketch of metric HP water hose fitting

All high-pressure pump units in Australia, Southeast Asia, and Europe using high-pressure fittings seem to have adopted either M24 x 1.5 or M22 x 1.5 as the thread of choice. In the main, M24 is the most popular. The 1.5 refers to the measurement from the top of one thread to the top of the next one—in this case, 1.5 mm. The ½" has been done away with, as it allows the insertion of unacceptable fittings not rated for the pressure, such as domestic plumbing joiners.

As stated earlier, by relying on the O-ring rather than the threads to make the seal, the joint can take a great deal more pressure. The metal surfaces need to be clean and SSG free. When you take a hose line apart, plug the ends or coil the hoses and join the ends; keep them clean. When you have a quiet moment, give them a squirt with WD-40 and clean them up a bit.

These threads do not need to be "strong armed". A simple little nip with a spanner is all that is required. (Some people do them up by hand only, but I do not recommend this.) Over tightening ruins the O-ring and screws up the mating faces. *Think.*

As previously stated, all HP jet O-rings are called 90 DURO. This is a hardness figure; standard O-rings are likely to extrude

Fittings Seals and Joints

out of the groove like toothpaste. It is impossible to crush 90 Duro by biting on it with your teeth, which is a good, quick field test to ensure you have the right hardness. Do not forget this when replacing O-rings. If you use standard O-rings, you will have failure after failure.

Autoclave Fittings
This is a nightmare fitting until you have it sorted out, and then you wonder why you ever used anything else. There are two or three styles of this fitting, the most common of which is the one made by Autoclave.

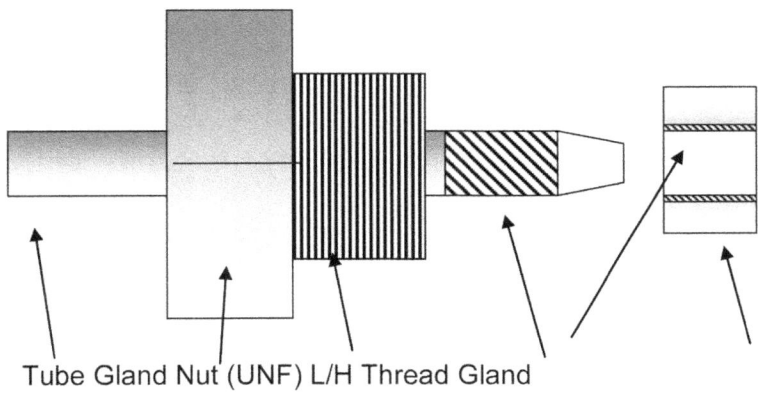

Tube Gland Nut (UNF) L/H Thread Gland

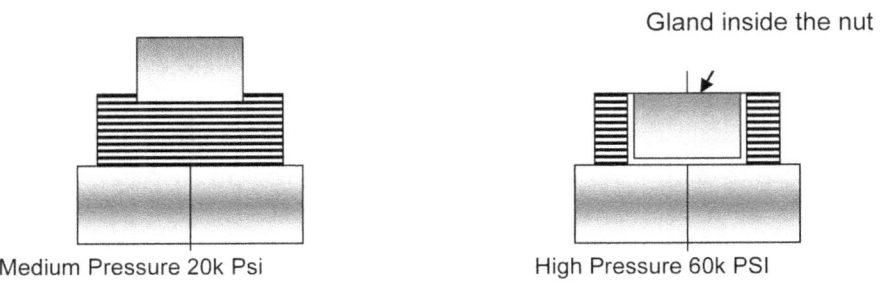

Medium Pressure 20k Psi   High Pressure 60k PSI

Gland inside the nut

Figure 9.5. Sketch of an Autoclave, or AE style, UHP fitting

There are two styles of fitting: AE medium pressure and AE high pressure, the difference being in the location of the gland. In AE medium (20k psi), as above, the gland nut presses directly onto the gland. In the AE high pressure (60k psi) fitting the gland is inside the gland nut. This feature contains the gland inside the gland nut, giving it extra strength. This is the only outward visible difference and can be used as a quick identification feature.

To identify a UHP fitting, measure the tube OD, *not* the nut thread. The dimensions will be in fractions of an inch. (Because it's not decimals, it's difficult to use a metric vernier. It will be ¼", 3/8", 9/16", and so on.). The most common in our industry is 9/16 for 20,000 psi or 60,000 psi. The hole through the middle dictates the pressure capacity of the tube. Except for the M14 version, most ultra-high tube comes from America, where they still operate in inches—hence the 9/16" measurement.

NOTE: The 9/16 gland nut will not fit the M14 metric tube and vice versa unless rethreaded to suit. The Europeans mark their gland nut; the Americans do not. The M14 gland nut, unless homemade, will normally be engraved with a 14. The tube size is the only size the dealers recognise, so if you have an Autoclave fitting with the instruction "Use ¼" Autoclave", it refers to the tube size. The gland nut will be something strange in UNC or UNF. Ignore the gland nut size; it will arrive the correct size. Do not forget to specify the pressure you want.

NOTE: Some European manufacturers thread the 9/16 tube M14 fine; this means that the gland must be M14 left hand. In *most* cases the gland is stamped "14" so you can see it easily. Be aware that this similarity can cause confusion; the two styles look the same but the left-hand threads will not be compatible and can cause a lot of grief in the field, especially if some idiot forces the nut. Insist on one or the other for all your hoses, and stick with it. I recommend you stick with 9/16", which is where most of the tooling available is.

To order, simply state 9/16" tube (or ¼", or what have you) gland and gland nut, and state the pressure. If you wish to fit a tube in place, you will need to rent the end-cone shaping tool and the left-hand die nut, which normally come as a kit. Alternatively the supplier will turn it up for you.

As with all face-to-face fittings, the Autoclave should not be over tightened, and the faces should be absolutely spotless. *YSE-approved (expensive) anti-seize* (Blue Goop or silver-based products), and apply it properly. If it develops a leak, you can re-cone the ends and reface the seat if you gall it. Because you did not use anti-seize, it could cost you many hundreds of dollars replacing the components that have welded themselves together forever. Using a *big* pipe wrench and gas torches will *not* help you; the fitting is ruined. "Sorry, boss, I didn't grease it. Just cost you 500 dollars".

Other Threads
Be aware that other threads do exist and in some cases may look the same but be quite wrong for each other. A good rule of thumb is that if they will not screw together two and a half turns by hand, something is wrong. Operators trying to force incompatible threads together have ruined many valuable components.

In the case of BSP and NPT, only ½" male and female will fit together. Even these two are a little different, and you should drop a compatible tap into the female to clean up the threads before trying to screw ½" together. The OD of the male NPT is 27/32", while the ½" BSP is 13/16". That's a difference of only 1/32" but just enough to gall stainless and spoil the fitting when you pull up tight.

PUTTING THREADS TOGETHER

O-ring fittings

- High and ultra-high pressure water fittings use O-rings pre-fixed 90 Durometer. These are considerably harder than standard O-rings; bite one and compare it. Soft O-rings will be extruded like toothpaste by the water pressure. Always use 90 Duro, even if you need to wait a day or two to get them. Soft O-rings encourage you to overtighten, stress the fitting, and damage the seating faces. And they will burst in a few minutes, so do not waste your time.

- Always lubricate the O-ring before tightening. Carry a can of silicone spray in your vehicle; a pinch WD-40 will do, and you *must* have a can of that in the truck. If using silicone grease from a tub, do not slap it about onto the O-ring; the surface simply needs to be shiny. If it is too thick, it will prevent the O-ring seating in its groove, and it will either be damaged when assembling the system or crushed when under pressure.

- Do not over tighten. The O-ring makes the seal and deforms to close the gap.

- Both faces must be clean.

- O-rings do flatten and become brittle, so check them out regularly. O-rings are cheap. And a leak always happens when the client is watching.

- Bottom style O-ring fittings are normally fitted with a lock nut. Nip down on the O-ring and crank down on the lock nut.

- *Never use thread tape or other sealant* on an O-ring or copper washer supported fitting. It does not need it, and you

- are only stretching the fitting by forcing it in. Get into the habit of using anti-seize on all your threads instead.

- Backup rings are always *after* or behind the O-ring when facing the pressure; the pressure pushes the O-ring onto the backup ring. It is a common fault to put these in back to front (O-ring behind), resulting in frequent O-ring failure. Lubricate the backup ring the same as you lubricate the O-ring. Backup rings are also available in 90 Duro; match these to the O-ring. If there is a problem getting rubber or buna backup rings of a suitable hardness, use Teflon split rings for backup.

Threaded Fittings

- Without fail, *always* use approved anti-seize when fitting stainless screwed components together. I cannot emphasise this enough. Don't use just any old anti-seize; use the correct anti-seize for stainless. It is much more expensive but heaps cheaper than a ruined fitting. If you can afford it get Blue Goop from the United States, it is the bee's knees and you will never gall a thread. The next best is a silver-based product; copper based is *not* good enough.

- The fitting should screw up at least six full turns for high-pressure applications. Here's a good yard stick: the thread should go in at least half its diameter.

- If the hex is smeared by a crescent wrench, change out the fitting.

- If someone has used a pipe wrench on the fitting, drop kick it. It is going to burst in your hand.

- Use the right fitting; most HP fittings are marked, and in most cases the working pressure is engraved on them.

- A female high pressure fitting should have a wall thickness of 6 mm or more. The male can have a little less but not much less.

- Do not forget that 316 stainless is not a strong material and should be restricted to a max pressure of 690 bar (10,000 psi). NOTE: 316 is nonmagnetic; go through your HP toolbox with a magnet and chuck out all nonmagnetic fittings.

- *Never use thread tape* on a steel high pressure fitting. Use Loctite 567 or 577 only. Thread tape will extrude. Loctite 577 will act as an anti-seize at a pinch, but you must coat both inside and outside.

And Finally...

Look after your threads, damn it. Protect them. They are fragile, believe it or not; one little ding can ruin a good fitting. I have seen stainless M24 joiners and other valuable fittings rattling around amongst the spanners in a toolbox, totally ruined because some moron did not have the brains to wrap a bit of tape around the threads and put them in a safe place. On clearing the toolbox, I estimated that it contained more than 800 dollars' worth of ruined mistreated fittings. They cost an arm and a leg to replace, and lost time screwing around with damaged fittings is not the sign of a professional operator. For every dollar spent on maintenance (fitting replacement), we need to earn five dollars. *Think.*

Invest in a *bottom* tap and a die nut for your threads, and clean them up when you have a spare moment. Always use a good-quality anti-seize on your threads. Make this a habit and be a pro.

Fittings Seals and Joints

If you really want to be a pro, go to a fitting supplier and get the little screw-on plastic thread protectors, available in all manner of thread sizes in both male and female. These are ideal for the job, reusable, and very professional. In Perth, West Australia, these are available from Specialty Plastics Pty. Ltd. Any money spent will be rewarded in extended fitting life and a reduction in "consumables" costs.

Chapter Eleven

Caring for Your Hoses

Hoses on reels must be put on under pressure. This is very important if you hope to get all the available pressure to your nozzle.

As a hose is pressurized, it inflates and the diameter increases. At the same time, it shrinks in length. If you have four, five, or six layers of hose on your reel, the bottom layer is going to be squeezed flat, if you are lucky. If you are unlucky, your hose reel is going to collapse. There is sufficient power there to crush the steel drum.

You cannot get the same flow out of an oval hose as you can out of a perfect circle. The circumference remains the same, but the available volume is greatly reduced. This is due to a phenomenon called laminar flow.

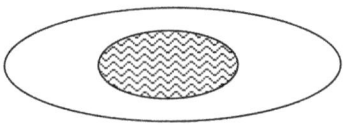 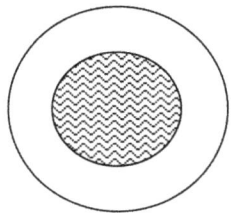

Figure 10.1. Laminar flow in a hose, pattern indicates full water flow

Water running through a pipe runs at differing speeds in different layers; the water around the wall of the pipe is almost stationary. This is due to friction on the pipe walls. Minute imperfections hold up the water. The next layer runs a little faster, sliding over the stationary water, the next layer runs faster, and so on to the centre. At the centre, a stream of water is running very fast. Each layer has a thickness, depending on the condition of the outer surface. If the pipe is out of round or oval, the stationary water at each narrow point is considerably thicker, causing a thickening of the stationary water, eventually leaving a very small bore of fast, free-moving water.

Laminar flow restrictions are what cause the hose friction losses to increase as a hose ages. As the imperfections on the outer wall increase in size, the layer of stationary water becomes thicker, slowing down the next layer, and so on. These imperfections are caused by damage due to excess bending, dirt in the water scouring the walls, and crushing during normal operations. The water can pick up tiny dags on the outer wall like three-cornered tears; these stand up in the water flow and cause "bumps" in the stationary layer. The water has to work harder to get past these bumps.

Squeeze the end of your garden hose between your thumb and finger. What happens? The squirt gets longer, and the velocity of the water increases. The pressure inside the hose increases, trying to get the water out. And all you really did was change the shape from round to oval. You still had the same overall circumference of hose at the opening.

"Turbulent flow" is another state water can get into while running through a hose. Turbulent flow is usually caused by the selection of too small a hose for the amount of water being pushed through it. Think of a smooth-flowing river; water moving smoothly along with not a ripple in sight (laminar flow)—until the banks close in and the water starts to rush. Waves and foam form, and the water becomes lumpy and starts to roar—we are now in turbulent flow. Energy is required to get the water through the restriction. The water becomes all mixed up and is constantly changing direction. Small eddies form, and the water is struggling to get out. That's friction at work. Friction robs us of the pressure we need to do our work; energy made by the engine, which is consuming fuel, is stripped off in the hose. The maximum efficient flow rate for any hose is 18m/sec.

MINIMUM BEND RADIUS
All hoses have a designed minimum bend radius. This is the smallest radius you can bend the hose without damaging it or weakening its pressure-carrying capacity or restricting its flow capacity. A good rule of thumb is a radius of 24 times its internal diameter. Check your supplier's literature for the minimum bend radius for a specific hose.

If you have a hose with a ½" nominal bore, the minimum bend radius is 24 x ½" or 12" or 300 mm. Do not get radius mixed up with diameter. Radius is the centre of a ring to the outside edge. Diameter is from outside to outside (the minimum bend diameter of the above hose will be 600 mm).

If you do bend it beyond its minimum bend radius, a number of things happen:

The high-speed central core of water running at velocity and turning a very sharp bend can punch through the slow and stationary laminates, strike the hose internal wall, and scour out the lining. This will either cause a burst or raise small dags at the scour point, greatly increasing friction losses.

The wire braid will open (try bending a hose restraining stocking and see what happens) and allow the internal lining to squeeze through and eventually burst. An indication of this having happened or about to happen is the appearance of bubbles of water under the outer cover at the bend point over the next few days.

When tying a hose off up the outside of a building, bring it up a bit higher than you need, tie a good double clove hitch to the hose and the handrail, and let the hose down until it is supported by the rope on the outside of the rail. If you have a spare hose stocking restrainer, this is the best way to hang a hose. But you will need to tape the end of the stocking to the hose. The hose will form its own bend radius over the rail as it pressures up. Do not use poly rope to tie off hoses, because it will cut the rubber outer cover; use hemp or similar.

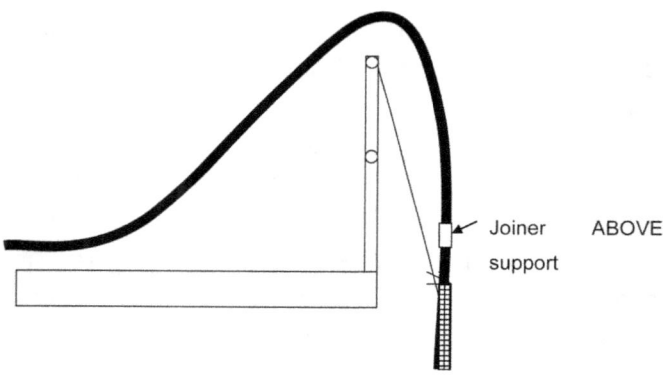

Figure 10.2. Tie off on the outside of the building

If you have a joiner in "mid-air", you must support the hose below the joiner. This may involve two ropes: one at the top and the other below the joiner. The joiner is not designed to hold the weight of the hose and the pressure. In most cases, the ends have a stocking; use the lower stocking, on the bottom hose, to tie off to. Allow the joint to hang in a loop in mid-air.

The rubber outer cover is put on to protect the braided wire reinforcement. It serves no real purpose in containing the pressure.

Caring for Your Hoses

If the outer cover becomes damaged, the hose is still good and will continue to hold design pressure—unless the braid rusts and a strand of wire breaks or is worn through. If that happens, the hose is unsafe to use.

Our hose standard allows us to operate at a safety factor of 2.5:1. That represents a design burst pressure of two and a half times the working pressure. All other hose applications—air, hydraulic, steam, etc.—have a minimum safety factor of 4:1 or more. We are permitted to use the reduced factor because of weight. If we had to use 4:1 hose, it would be so heavy and so stiff we could not move it nor pick it up.

Because of this reduced safety factor, our hose inspection and test procedures are more stringent. We *have* to inspect our hose before we use it each day. One broken braid out of a typical four reduces the safety factor by at least 25 per cent—too close for safety. Therefore one broken wire means the hose has to be repaired and cannot be used until then.

HOSE INSPECTION

Each day, before going to work, visually check your hose for damage. Look for the following:

- Look for broken wire braids, normally seen by looking at points of damaged outer covers.

- Check for bubbles in the outer cover, indicating a leak through the inner lining.

- Watch for movement of the ferule off the hose, indicated by a small ring of clean hose just behind the ferule.

- Check for ferule cracking by looking for them at the lines left by the ferule crimping machine.

- Look at the back of the hose nut (if used). It should still be square to the nut and not dished, indicating it has been grossly over tightened.

- Check that the nut is still in good condition and that the corners are not badly smeared by incorrectly adjusted spanners.

- Check that the nut is not cracked along its length. This is normally caused by dropping a hose from a height and the nut striking the floor at speed, accelerated by the weight of the hose.

- Look inside the nut and check the threads for wear. Compare these to the threads deep inside the nut; these never get used and are "as new".

- Check the globe seal, the hose tail to joiner connection inside the nut. Look at the O-ring; it should be in good condition and not flattened and chipped. Replace it if it is suspect. Wipe the threads and the globe, if you have it; a squirt of WD-40 would not go amiss. Do not use thread tape or sealant on these threads. They do not hold water; they simply keep the face and the O-ring together. If they leak, the O-ring is shot. If the joiner is stainless, use standard anti-seize on the threads. Only very special and expensive hoses have stainless hose nuts, so stainless-to-stainless anti-seize is not required. TIP: Always carry two or three O-rings on your key ring; that way you always have one when needed.

- Run your eye along the hose. There must not be any dog legs or kinks in it. And it should not be flattened (where the truck drove over it yesterday). If any of those conditions exist, tag it out of service and have it tested. Dog legs and kinks are the points where the hose is going to fail. If you are hand holding the hose, it is extremely important that you not use it, not even to "just finish this job".

HOSE PRESSURE TESTING (Standard AS/NZS 4233)

A new hose must arrive in your yard with a test certificate showing that the made-up hose and ends were tested at the time of assembly. The hose will be numbered (usually provided by you), normally engraved or stamped on the hose nut on one or both ends. This test certificate should be stored in the hose maintenance log. (Haven't got one? Get one; it is a legal requirement.)

A repaired hose must also be pressure tested before it is returned to service. This test is certificated, and you should cite the test certificate. If the numbered hose end has been replaced, the numbers should be transferred to the new hose tail or nut.

The certificate will say something like "This hose, No. XYZ, has been tested for 5 minutes at XXX pressure and at the time of the test is certified safe to use". The test does not imply that it will be safe to use next week. The tester has no control over what you do to the hose once it leaves his test bay. The test certificate is not a guarantee, although most reputable hose assemblers will honour a warranty claim if a hose fails prematurely.

Note the following hose killers:
- Bend radius (Pay particular attention to where the hose is joined to the pump head. Fit a rope support to ensure the bend load of the pump head to ground level is not on the ferule). The radius of any bend in any hose should be restricted to twenty four times the inside diameter: 1" hose equals 24" radius.

- Floor grating

    Hot pipes

- Pipe welds

- Pipe threads

- Flanges

- Stairs

- Steel columns

- Vehicles driving over them

- Concrete floors

- Dirty water

PIRANA OR SIMILAR DRAIN HOSE

These hoses do not have wire braids, and they are typically nylon or Kevlar. As a result, they do not have good body strength and can twist and become easily damaged. The reason for the lack of wire braid is to keep the weight down to allow you to travel long distances up pipelines.

It is absolutely critical that the lower layers/coils of this hose are installed on the drum under pressure. This hose is so soft that it can be easily flattened and closed off completely. Always draw the hose back onto the reel under pressure (this applies to all hoses on reels), at least until the final or outside layer.

To do this, fill the hose with water, get all the air out, and plug the end. Back off the regulator to zero pressure. Start the engine and run at about half speed. Crank down on the regulator, and watch the gauge until you get about three-quarters of normal working pressure. Engage the hydraulics and begin to wind the hose onto the reel; this is a two-man job. Carefully align the coils as you wind on. Once completed, back off the regulator and shut down the engine. Once the pressure is off, the hose will relax a bit and may look a little untidy. Don't worry; as soon as you pressure up again, it will neaten up once more.

Caring for Your Hoses

These hoses must be used carefully. Normally a reel may contain as much as 600 feet (standard coil) or 200 metres. The cost of a coil is between four and five thousand dollars. You can destroy it in five minutes. A join on your hose reel looks awful, damages the other hose, and is a friction maker.

Never allow the hose to become kinked. If you see a loop of hose on the ground, do not simply pull it out; it will not come, and the hose will permanently kink. At this point, the hose will eventually burst. Untwist the loop most carefully by hand, and guide it into the pipe or onto the reel until the hose is straight again.

The outer cover cuts very easily. A pebble in a concrete manhole cover can slit the outer cover like a knife. Always guide the hose around corners through a sacrificial pipe. The best stuff for this is plastic agricultural irrigation pipe (ag pipe), used by farmers and the like. It is normally sold in a coil, and a half loop is just right to pass your hose through. Drill a couple of holes in one end, and tie a piece of support rope through it. Slide your drain hose through, fit the bomb, and lower it into the hole. Allow about half the ag pipe to enter the hole, and tie off the support rope to your truck.

Get just the right size—just right for the hose and too small for the bomb. Push the hose through, and then attach the bomb. Now you have a device which lets you know that the bomb is home before it comes out of the drain. The ag pipe never needs to come off and can be stored on the hose on the reel.

It is a good idea to mark your hose about 6 then 3 then 2 metres from the end; this flags where the bomb is when you are pulling back and on the way out. The best thing for marking the hose is adhesive electrical shrink tube. It comes in all kinds of colours and, once shrunk on, it will outlast the hose.

HOSE AND FITTING PRESTART INSPECTION

All water blast and high pressure hoses must be inspected prior to use. This is a standards (AS/NZS4233) requirement and is an important part of your prestart checklist. I repeat myself here, as this is a most important activity; your life or continued good health may depend on it.

Look for the following:

- A bulge spot on any of the steel fittings; this could indicate that cavitation has eaten through the inside and the hose is ready to burst. Smearing of the hexes caused by ill fitting wrenches, cracks or damage from tools with teeth. Replace it.

- Any movement of hose off the ferule: look for a ring of clean or new hose just behind the crimped ferule—or, if it's a skived hose, look for wire braid. If one of these indicators is present, remove the hose from service and send for repair.

- Any evidence of splitting of the ferule: look for a crack along the lines left by the crimper. Remove hose from service and send for repair.

- Inspect the hose nuts for cracks or grooves caused by pipe wrenches, and tag out the hose if either is found. Seriously smeared nuts and fitting caused by badly adjusted crescent wrenches should also give you concern. Remove the hose from service and send for repair.

- Check that the nut's rear metal work is not bowed from over tightening; wiggle the nut on the hose tail and check for excessive play. Make a note of how a new one feels. Plan to replace the hose end ASAP if it is too loose in your opinion.

- Inspect the seats of the joiners you intend to use; wipe them out and check for damage on the faces. Replace or repair if damaged.

- Inspect the O-rings inside the hose nuts; replace if flattened.

- Inspect the whole hose for broken braid; discard it if one wire is broken. Remove the hose from service tag out and send for repair.

- Inspect the hose for obvious damage: kinks, crush, and surface bubbles indicate a fault that requires the hose to be taken out of service and repaired.

As with all things related to high pressure water jetting, hoses should not be run at their manufacturer's maximum working pressure if you want economical life. If possible, always run about 20 per cent under maximum working pressure.

All equipment damage is an "accident". (You didn't do it on purpose, did you?) As such, all equipment injury or damage should be reported as you would an accident or injury to a person. It should also be investigated that way. The outcome of that investigation should be the creation of policies or procedures to ensure it does not happen again.

Hose is the most expensive component of operating costs (consumables). A typical pump unit will make a profit of about 500 dollars per day after all costs are taken out. This profit is put towards building the company, buying new plant and equipment, and providing you with job security. You can blow that in a few seconds of inattentiveness, carelessness, or clumsiness. It may mean little to you today; you will still get paid on Friday. However, you should ask yourself if there will still be a job for you next year? A 20 metre length of 1000 bar hose will cost about 1,300 dollars; a 200 metre length of Pirahna will cost in the region of 4,000. How many days of working at cost will it take to replace it? Two extra minutes in correctly placing the hose or fitting a wear strip, tiger tail or tube is all it would have taken. Why didn't you do it?

You may drive out of the yard with equipment worth in excess of half a million dollars. Your boss most likely mortgaged his house to buy it—his, not yours. Do the right thing and reward his investment by taking care of the equipment as if it were your own. Your boss has paid you a compliment in trusting you with his family's home. Repay that trust by thinking.

Chapter Twelve

Tools You Can Make, Guns and Gadgets

WATER LIFTS
Tremendous water lift power can be obtained using the water jet. There are a number of units on the market, off the shelf, that are made to lift waste and sludge. Making your own is really quite simple and fun too.

Drill a 25 mm hole into a long 4" bend. Locate the drill bit so that it is directly facing the centre of one of the open ends.

Screw a ¾" socket onto a piece of ¾ inch pipe. Pass it through the hole from the outside, pulling the socket close to the outer edge of the drilled bend. You will need to grind the socket off to suit the curvature of the bend. Use the piece of pipe as a guide to ensure the fitting is in the centre of the bend outlet.

Weld the socket home.

Take a solid 1" bar stock thread it ¾" BSP, and drill the centre out to suit a 1/4" nozzle on one end and a male fitting to suit your hoses on the other. Taper the hole out so the water flow runs into the nozzle with the minimum of turbulence.

Fit your nozzle, and screw the bar stock nipple into the welded-in socket so that the nozzle is pointing out of the open end of the bend. The inside of the bend needs to be smooth; the nozzle should not stick out into the flow of spoil.

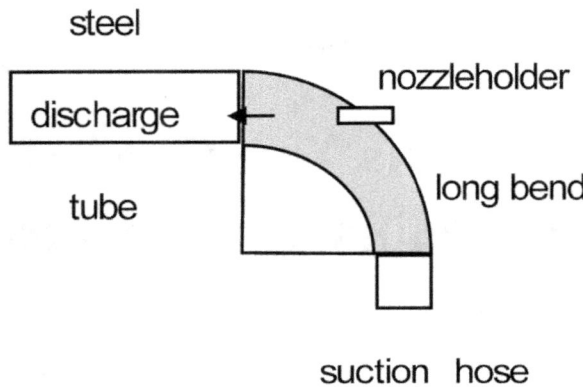

Figure 10.3. Water lift

Fit a 2 metre length of steel pipe to the end with the jet spraying through, and fit a discharge hose to that.

On the other end of the elbow, fit a suction hose or hard pipe.

Attach your water jet hose to the other end of the nozzle nipple, and engage the pump.

You will be able to lift 75 mm ball mill balls (cannon balls), nuts, bolts, and huge amounts of sludge.

The trick with water lifts is that you need some airflow or you cannot create a vacuum. If you intend to stick the suction end into a solid pool of mud or sludge, you need to make a vent. To do this you need to place a 1" ball valve onto the side of your suction pipe at least 1 metre below the bend. Open this valve fully, start the pump, and as soon as you have product flow, slowly close the valve. I fitted an elbow down the bottom end, ran a length of 1" hose up to the bend, and fitted a ball valve there; this gave me my vent with its control valve on the surface.

This is a great tool for lifting sludge out of a manhole after drawing back with the drain cleaner. Fit a separate hose and tee off this supply. Once you have drawn back the dirt, flip your valve and suck up what you have drawn back, and discharge into a tank or bin. Once clear, throw the valve the other way, and carry on drawing back. This eliminates the need for a vac truck.

AIR LIFTS

Air lifts are great tools for some jobs. For them to work properly, you need at least two times expansion of the air bubble. They will work with less, but for spectacular results 2 atms is best. A balloon filled with air halves its diameter every atmosphere it goes under water. To get two times expansion, we need a depth of water (height of lift) of 20 metres. So, to clear solids from 20 metres of water depth and deeper, an air lift is the way to go. They work as low as 5 metres, but they are not as effective as water lifts at these depths. As the air expands, rushing to the surface, it acts like a piston drawing fluids and solids with it. If you are working in shallow water, you can cheat a bit by adding height to the outlet pipe; once filled, it will provide a similar situation to water head.

We recently emptied 600 diameter tubular piles 40 metres deep, filled with water. The 4" air lift took every drop of water out in seconds.

You can also fit a non return valve to the bottom of your pipe. Stand it in shallow water with the pipe supported overhead to give the lift height, fill it up to the top with a hose, and start the lift. It will work nicely. The more height the better it will work.

High Pressure Water Jetting – An Operator's Manual

Take a piece of light pipe about 3" in diameter and 2 metres long, and drill 60 2 mm holes in a ring all around the pipe about 600 mm from one end. Make up two big washers with >3" centres, to suit the pipe OD, and about 5" outer diameter. Weld these onto the 3" pipe above and below your ring of holes. Wrap a piece of plate around the outside of the flanges, forming a box around the drilled holes.

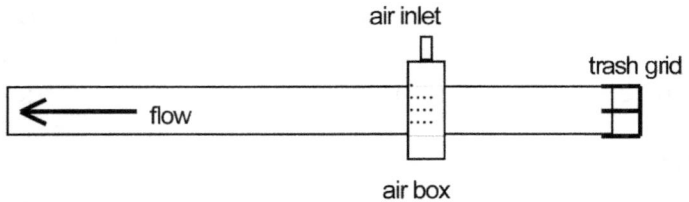

Figure 10.4. Air lift

Drill a 1" hole into this outer plate, and weld on a suitable socket and air fitting. Bend two pieces of 5 mm round bar into a U, and weld them onto the end closest to the box as a X grill; this is now the suction end. This grill is to prevent big pieces of solids entering and, if you are working the nozzle end by hand, to keep your foot out of it.

To the other end fit a length of grey Heliflex or similar 75 mm vacuum hose. It must be substantial, as the vacuum generated by the expanding 2 mm bubbles will create an incredible lift. As a diver using a 4" air lift made as above, without the grill, in 80 feet of sea water, I managed to crush and suck up a hard hat. Sucking a good vacuum hose flat is little trouble. The small bubble formation as described will outperform a single-hole entry by 100 per cent. You will need up to 100 cfm of compressed air at about 80 psi. Ideally you will need some method to break the vacuum if you get a blockage. This could be a 1" ball valve fitted into the wall of your hard pipe just above your air box.

CLEAN AROUND CORNERS

Quite often you need to clear degraded concrete from the underside of the lid of a manhole before you can enter. This is very difficult thing to do from the outside. Make up a tee piece to fit onto the end of your gun barrel or onto a length of hard pipe that will fit onto your hose end. The tee needs to be a "street" tee—a tee with different sized holes in it. If you have the tools, make this up out of a solid piece of stainless.

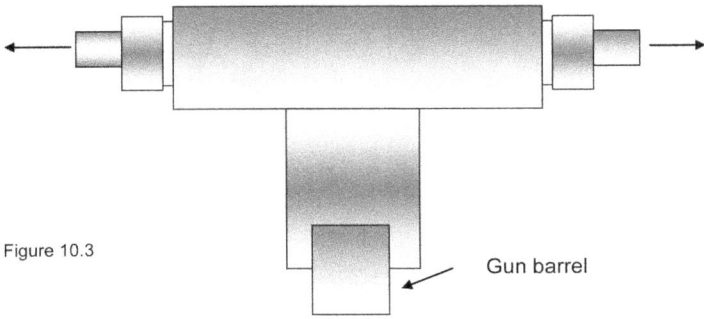

Figure 10.3

Gun barrel

Figure 10.5. Tool for cleaning tough spots

The hole pointing down, as above, will be threaded to suit your gun barrel or rigid lance, and the two side holes will be ¼" NPT or BSPT, depending on the nozzles you have. They must be exactly the same size and flow rate. NOTE: Nozzles of the same number can differ in performance. Visually match a pair by running at tap pressure and watching the flow.

Do not forget that the material you use *must* have a 2.5:1 working over burst ratio.

This is a most useful tool for any drain work; it has no reaction force, so it requires little effort to use, and you can fit as big a nozzle (2) as you like. This gadget can be used to clean almost anywhere.

To clean the underside of the manhole lid, simply pass it in and tilt the gun. Walk around the hole, jiggling the holder about to clean the underside of the lid.

Always test the nozzles at tap pressure before applying pump pressure. Do not engage pressure while the tee is out of the hole or uncontained. If a nozzle is blocked or becomes blocked, the one firing nozzle will spin you around like a top; you will not be able to hold it.

JOINING TWO PUMPS TOGETHER

It is very useful to be able to join two or more pumps together to give added flow to provide the reactive force required to run your nozzle a great distance along a horizontal pipe or up a vertical one like a chimney stack. The best I've done was a 66 m high flare stack in a petroleum refinery (quite impressive when the nozzle came out the top). Once you know your hose weights and have mastered reaction force calculations, you can put your nozzle just about anywhere.

The two pumps do not have to be of the same flow rate, but they should be capable of as near as possible the same output pressure.

Select your nozzles for the total output of both pumps.

From each pump, run a 20 m hose to your Y piece splitter block. (*Do not use a tee; there will be huge friction losses.*) Ideally your splitter block will be ported ½" x ½" x ¾". It is very important that there is a full hose length between the splitter block and each of the pumps. If they are too close, the regulators will start to fight each other and cause surges to occur—and possibly blow a burst disc.

Fit your nozzle holder to the end of your hose. The nozzles fitted must be capable of discharging almost the total flow from each pump.

Fully back off both pumps regulators to zero pressure. Start both pumps, and bring engine revs up to running speed. Fully shut/close the regulator on one pump (pump 1) to that max pressure setting position.

Closely watching the pressure gauge on pump 2, and begin to close off the regulator on that pump. Ignore pump 1. Pump 2 is the pressure regulating pump for this operation. Once the system has reached operating pressure, leave everything alone. Don't fiddle.

MAGNETS

Magnets are an incredibly useful part of your tool kit; they come in all manner of shapes and sizes. One of the most useful is the Bear Paw magnet that has an on/off lever and is fitted with an eye to tie things too. You can get them with a holding capacity of up to 200 kg (440 lbs). That can hold a lot of reactive force or even you at a pinch. They can work underwater too. I have made caterpillar-tracked machines with magnets on the grouser plates to climb up tank walls.

Electrically controlled magnets can be switched on or off. However, using electrical equipment around water is always risky.

The capacity of a magnet has a lot to do with the cleanliness and thickness of the steel to which it is attached, so *always* test first. Ten to 12 mm (1/2") plate is fine.

Using magnets to mount jigs is a much better and quicker method than welding or messing about with scaffold. Some of the small, 50 mm (2") rare-earth magnets will hold a direct pull of 25 kg or 50 lbs. When fitting magnets, keep your fingers out of the way. Get a finger trapped between the magnet and the steel, and say good-bye to your finger.

Always store live magnets with keeper plates on them, and liberally squirt WD-40 over them when not using them.

USING A WATER JET HANDGUN

The water jet handgun is a safety device. It provides the operator with a means of controlling the water flow out of the nozzle. It can be turned on or off as needed. If we did not need that control, we would fit the nozzle to the end of a piece of tube and weld some handholds onto it. In the process, we would eliminate the friction loss, typically about 7 per cent, and save ourselves more than 3,000 dollars, the cost of a typical gun.

Any person hand-holding a lance or gun *must* have a means of instant shutoff, be it a gun trigger or foot control valve. Any person hand-holding any lance or gun *must* be observed by a trained observer (must be able to see the operator at all times), who has a means of instantly shutting off the water supply to the nozzle in the event the operator moves into danger.

However, the operator needs to be able to control the water flow to ensure his safety and, in the event that something goes wrong, it provides a form of "dead man" switch to ensure that if the gun is dropped or the operator slips, the flow is immediately shut off, eliminating a potentially serious hazard.

There are two types of handgun: the wet dump and the dry shutoff.

The wet dump gun dumps all the system pressure to atmosphere when the trigger is released. This is normally via an open-ended second barrel on top of the gun. With computer pressure-controlled pumps, wet dump guns cannot normally be used. The computer sensing a loss of pressure either speeds up the engine trying to maintain pressure or shuts the unit down due to too much speed or loss of pressure situation. If the computer is monitoring revs only, then maybe a wet dump gun can be used. Pumps without computers or regulators can be used for wet dump guns.

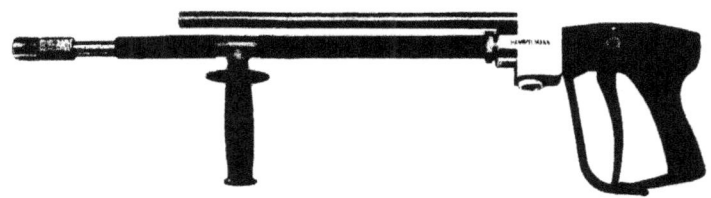

Figure 10.6. Hammelmann wet dump gun (note the dump barrel)

The dry shutoff gun shuts off the flow to the nozzle at the gun, maintaining full pressure in the system between the trigger and the pump. The regulator, sensing a rise in pressure, opens and allows excess water to discharge back to a tank via the regulator dump port, whilst maintaining system pressure. Once the trigger is pulled, a pressure drop is indicated, and the regulator begins to close until the set pressure is leaving the nozzle.

If the nozzle is slightly smaller than needed, some water will be allowed to pass through the regulator all the time. This is not a good practice. Not only is wear being generated on the regulator (overhaul kit 1800 dollars), but you are also wasting diesel fuel just to pump water back into the tank. Every litre of water costs you, so 20 per cent of your water going back into tank is a lot of fuel wasted.

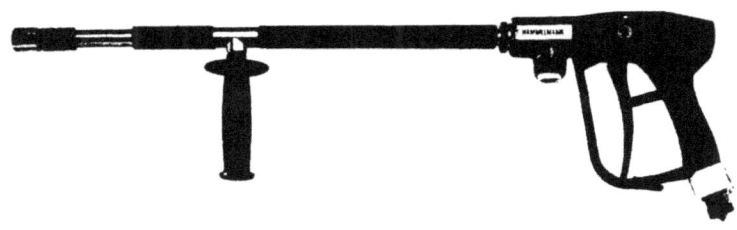

Figure 10.7. Hammelmann dry shutoff gun

The picture above is of a dry shutoff/electric shutoff gun with the electrical connection attached to the bottom of the trigger handle. This electrical connection on a dry shutoff gun is an option.

Both guns pictured show an unacceptable situation. Can you see what it is?

The Australian Standard AS/NZS 4233 states that the nozzle should be at least 1 metre from the trigger. In exceptional circumstances, a gun may be used with a shorter barrel but, section 6.7 states,

> (f) where practical the length* of the gun barrel should be such that the nozzle strikes the ground before the operator can inadvertently direct it onto his feet or legs. NOTE: it is recognised that in some circumstances it may be necessary to use a much shorter gun barrel on the gun, at such times extreme caution must be exercised on the part of the gun operator and other team members....
>
> *This length has been accepted as not less than 1 metre for Australian operations. US is 48".

Ninety per cent of water jet injuries are to the foot— typically from behind the steel toecap up to the ankle. It follows that metatarsal protectors should be worn when using short gun barrels. If you feel metatarsal protectors are uncomfortable and impractical to wear, *do not use short gun barrels.*

Let's return to the standard:

> <u>2.2 Limitations and Use</u> High pressure water jetting systems should not be used unless the operator is satisfied that:
>
> 2.2.a   It has been inspected or serviced in accordance with the manufacturer's requirements

2.2.b   It is free from any defect that has been identified at the last inspection or service.

To the best of your knowledge, the gun is in good working order, is safe to use, and its safety devices are in a safe and operational condition, yes? Go to work. A "maybe" or a "not sure" means that the gun stays right where it is until you are sure. Remember the OHS Act. If you take it out knowing it had a fault, you are exposing yourself to a fine *and* an injury. Both of these documents will condemn you.

You must check that the gun is in good working order and safe to use before you use it. Hammelmann dry shutoff guns can fail in the on mode. This is quite scary; you release the trigger, and gun continues to fire. Now what do you do? If the foot pedal observer is on the ball and watching you, he can get his foot off the pedal and drop the pressure. If he is not, you are in trouble.

This failure is not a sudden happening; you will have been given signs. One sign is a water leak out of the gun body above the trigger; the other is water dribbling out of the nozzle when the trigger is released. Did you inspect it before you started work? Why not? Got a problem? Tag it out, and send it for repair. Do not take a chance "she'll *not* be right".

Some guns have a joiner halfway along the barrel where extensions can be screwed on. This joiner has an O-ring in it and seals well—as long as the extension was not dropped on its end or stored upright in the truck. A small ding on the O- ring groove means that it will come out and a leak will occur just where you have your hand. Water jet injuries to the hand are extremely painful and require a trip to hospital. *Please* carefully inspect the gun and all its parts before using it. If in doubt, do not use it.

The trigger must *not* be jammed or tied in the ON mode. If you drop it or fall, the trigger will not automatically release, and the water jet will strike you—maybe more than once. If your hand is getting tired, swap out with the observer and let him or her have a go. Jamming the trigger is "Tampering with a Safety Device". OH&S has a number of regulations they can use if you get caught.

The nozzle holder is not a hammer. If you have a hard piece of scale that will not come off with a water jet and it needs persuading, use a hammer or chisel to remove it. Jabbing at it with the nozzle holder puts all manner of strain on the threaded fittings, which may suddenly start to leak and make a hole in you.

Upon completion of your work, wash out the gun with clean water. Pay particular attention to the area where the trigger goes up into the handle. Air dry or wipe the gun down and *hang it up* off the floor. Squirt a little WD-40 up into the body of the gun at least once per week.

The gun is a tool of your trade; look after it.

BEFORE STARTING WORK

Check out the gun/foot pedal visually, and pull the trigger and release it. Does the trigger snap smartly back to the OFF position? If not, there is something wrong. Sort it out by doing the following:

- Remove the nozzle holder cap and pulling out the nozzle. Check the condition of the nozzle and O-rings. Do not any replace yet.

- Connect the gun to the hose, pull on the trigger, and allow water to run through at hydrant pressure for at least thirty seconds, flushing the system.

- For a dry shutoff gun, operate the trigger to ensure that it closes and that the spring pressure is adequate. There should be no water flowing out of the nozzle.

- For a wet dump gun, pull the trigger and ensure that no water is coming out of the dump tube. If the dump tube continues to run with the trigger pulled at low pressure, at high pressure you are going to lose most of your flow to dump. If it leaks, get it serviced.

- Refit the nozzle, check that the backup washer is up against the step, and lightly lubricate and fit the O-ring.

- Insert the nozzle into the holder by pushing with your thumb; you do not need to hammer it in.

- Fit the nozzle holder cap and hand tighten firmly. You do not need to use a spanner or over tighten any component that is sealed with an O-ring.

- Turn on hydrant flow once again, and pull the trigger. Inspect the flow out of the nozzle. This should be a "glass" rod about 250 to 400 mm long before it starts to feather. If the water stream starts to break up just as it leaves the nozzle, the nozzle should be replaced. Your nozzle is your scalpel blade; blunt blades do not work.

- Apply half pressure to the gun and check for leaks with the trigger OFF. Attend to any leaks with a dry shutoff gun or trigger pulled ON with a wet dump gun.

- Pull the trigger and check for leaks out of the body and the connections. Attend to leaks. NOTE: Wet dump guns should not run water through dump tube when working at pressure with the trigger pulled. Deliberately allowing a dump gun to leak by not pulling the trigger fully on will rapidly destroy the valve in the gun.

STEPS FOR SETTING UP ON THE JOB

- Plan your moves and action sequences.

- Take 2 minutes to check for unforeseen hazards.

- Before starting the pump:

  o Make sure the nearest eyewash station works. Flush out the nozzles and ensure that they will do what you expect. Clear any debris or obstructions between your workplace and the eyewash. If you need it, you will most likely have your eyes closed.

  o Check your workplace footing. Clear away debris from where you will be standing.

  o Check your hose and make sure you have some hose slack behind you and that the hose cannot fall off whatever you are working off. If it suddenly wrenches onto you, it could pull you over or cause you to lose control of the gun. If necessary, tie it off with a piece of rope.

- If working from a scaffold, make sure you have checked the Scafftag and that the scaffold was erected for water jetting work. Your reactive force could push it over. Check the following on the ladder or scaffold:

  o Ladder tied off and entry gap closed to prevent you falling through

  o Minimum of three boards tied off securely

  o Kickboard secured

  o Midrail secured

  o Top rail secured

  o Gap in front of your footing is less than 100 mm (If a hose blows, you will certainly take at least one step forward to recover your balance; that step should not be into space between the job and the front board.)

- o Scaffold is back-stayed or secured to the structure being cleaned

- Check to make sure your observer can see you and that you have clearly established a method of signalling your intentions or needs.

- Check the line along your water jet splash and spray direction; ensure proper barricading and check for other trades above or below you.

- Advise other trades that you are about to start work.

- Make sure you have the correct PPE for the job you are about to do and are wearing it.

Place your feet in a braced position, grip the gun firmly, and start work

It is always good practice to start working at a reduced pressure until you are quite comfortable, once satisfied get the pressure increased to a comfortable pressure that suits your body build and strength. Quite often the job can be completed as well at 1000 bar as it can be at 1200 with less load on you.

Chapter Thirteen

Confined Space Entry

These notes are for discussion purposes only and are not considered suitable as training in confined space entry. All persons entering a confined space of any manner *must* be formally trained, certificated, and authorised to do so.

These notes are provided as a safe working practice for employees working in tanks and vessels, municipal and council manholes, drains, and sewers. The information provided is accurate to the best of the author's belief, but is offered without warranty, expressed or implied. Additional information may be needed in some areas with regard to unusual or special applications. Contact the owner of the sewer or drain for any local conditions or restrictions that may not have been addressed in the following pages.

The information provided should not cancel or lessen the instructions given to you by your employer or the owner of the sewer, drain, or manhole. Be guided rather by these persons, as they will have better knowledge of the local conditions. By all means, discuss their instructions with them, bearing in mind the information in this chapter.

Where used, the word *should* indicates a recommendation and the word *shall* indicates an essential practice. The places where the term *trained* is inserted indicate just that. However, if no formal training is available locally, the operator shall be instructed by a competent and experienced person and shall have practiced the actions outside the job place.

All projects carried out in any confined space, sewer, drain, or manhole for or on behalf of a client should be supervised on site by a trained Safety Supervisor. This person shall be designated "the responsible person".

The objective of this chapter is to provide a grounding of safe working practices for operators working in confined spaces with special reference to sewers and drains.

The Health and Safety in Employment Act 1992 dictates that all significant hazards are identified and addressed to ensure a safe place to work. This chapter aims to provide workers with some of the basic knowledge to work in a confined space and a hostile atmosphere.

The definition of a confined space includes but is not limited to:

1. any work space which has only one entry or exit; or

2. any space not normally accessed as a work space; or

3. any work space which has no direct ventilation (or requires mechanical means to ventilate) and an atmosphere that may contain substances hazardous to health; or

4. any tank, vessel, tunnel, or structure that requires entry or exit by way of hoists, ladders, or scaffold with a depth greater than two metres; or

5. any trench or culvert with vertical walls of a depth greater than two metres.

Interesting note: A recent case was brought against two workers who did thickness testing on *top* of a flat storage tank 20 metres across. This was classified as a confined space—there was only one entry of exit, the area was not designated as a work space, and they were unsupervised. They also were not wearing fall arrest gear. I researched this a little and found that you can be penalized for standing on top of a shipping container without a harness and secured tag line if that container does not have handrails and a stairway access. (Sky hooks?)

IMPORTANT NOTE:
A confined space is considered entered when you have or may have placed your head and shoulders into a space as described above.

Any accident or injury sustained for which you may be prosecuted under the HASE Act 1992 may be as a result of a contravention of one or another standard, code of practice, or regulation. It is therefore advisable to be aware of regulations, codes of practice, and standard requirements. The primary requirement—as stated in the HASE Act, The Duty of Care—is to provide a safe place of work.

SPACE ENTRY DISCUSSION
Entry into or through a vessel, bail fill, sewer, or storm water manhole is extremely hazardous. Most of us look at the manhole plate set in the road and have little or no idea of what happens below it. In some cases, some form of lettering is cast into the lid which sometimes indicates what particular service the lid covers. This sounds a little vague; it is meant to. Always assume the worst case, regardless of what you understand or have been told. If there is a way to do the job without putting a man up the pipe or down the hole, use it.

SEWERS
Sewers are designed to carry off human waste, bath and toilet flush water, water, detergents, etc., and a certain amount of

toilet paper. Most users, unaware of these limitations, thoughtlessly dump disposable nappies, sanitary towels, domestic cooking fats, and all manner of waste for which the original designer did not allow. Where the sewer serves industrial installations, unscrupulous users often discharge chemicals, poisons, and other toxic materials into the system. In many cases, the waste deposited is not necessarily toxic and the culprit is under the impression that what he is tipping into the sewer can't hurt anyone. Though two persons are discharging "can't hurt anyone" waste, the combination of the two chemicals may easily become lethal, either giving off a gas or as a liquid which may splash onto your skin and hurt you.

Gas is constantly being generated in sewers. It *will* be there. The breaking down of human waste generates hydrogen sulphide (H_2S) which can, in sufficient concentrations, kill you in a matter of minutes. Swimming pool chlorine flushed into the sewer can create a lethal gas in the pipe. Bleach will do the same if sufficiently concentrated. Ammonia is another common household chemical that is washed down the drain. It too can kill if concentrated.

Oxygen is consumed by the breaking down of human waste, and the atmosphere in a sewer is often either low in life sustaining oxygen or totally without it. There will be something there. This is not a vacuum; your lungs will move in and out. But whatever is in that air may well kill you in a matter of minutes.

Many viruses and germs thrive in sewers. Any open wound, even a tiny scratch, can fester and go seriously septic rapidly in this environment. In Australia, I could not find a recorded case of a sewer worker catching hepatitis from working in a sewer. Leptospirosis is not uncommon; this is caused by an open wound coming in contact with rat urine. You must practice good personal hygiene at all times and wash your hands in a goo-quality antiseptic before you use your hands to eat or smoke. When working in sewers, always wear waterproof gloves. Contrary to popular belief, there is no chance of you getting AIDS or another sexually transmitted disease from anything you may encounter in a sewer.

All cigarettes must be extinguished within 4 metres of the open manhole. Naked flame must be kept at least 6 metres from the open manhole. Gases which may rise from the manhole can explode when ignited.

Practice strict safe working habits, and never take anything for granted. Ventilate all manholes before you enter, and do not ever enter a space unsupervised. Unventilated contaminated sewers require that the operator wear a breathing apparatus either from a compressed air bottle he carries or use surface-supplied air via a hose. Entry into 99 per cent of sewers or drains can be carried out without the use of breathing equipment when using approved air movers. Ventilation is always possible with a little effort and forward planning.

A man unencumbered by bottles, masks, and hoses will be ten times more effective. Not only is the added weight a problem; having to draw your air through a regulator, breath out through a discharge valve, operate with restricted vision, and work fitted with a rubber mask all add up to a massive reduction in work capacity.

Always wear a full-body, approved safety harness and always have a secured recovery line between you and your supervisor. The end of the safety line must be tied off to a strong point on the surface. Keep in contact with each other by talking; this way the supervisor can hear that you are okay, even if he can't see you.

If you are not able to pull an unconscious man out of the manhole by hand, you must have either a second man or an approved man lifter and an A frame that could be used in an emergency. This *must* be set up over the hole and not in the truck. If you need it in a hurry, it will not be there. And as sure as nuts, some vital part will be missing.

If the man in the hole collapses for whatever reason and you need to go in to assist him—*first* get help, and then affect the rescue. If you don't, there will be two unconscious or dead men in the sewer.

A recent death involved a bricklayer entering a brand-new manhole to plaster the benching without supervision. He lifted the lid and climbed down two metres. At lunchtime, he was missed, and his mates went to look for him. He was dead in the bottom of the hole. The drying cement plaster placed the previous week had absorbed the oxygen, and he died of oxygen starvation. It's really easy to die in a confined space.

STORM WATER DRAINS

Storm water drains are designed to carry away rain water. They suffer from misuse almost as badly as sewers. Dog droppings, carwash waste, grass clippings, illegal waste disposal, and bad or leaking sewers can all contribute to creating a noxious atmosphere in the line. Treat all storm water drains as sewers, and you won't go far wrong.

Most drains are open to atmosphere via the drains every few metres along in road and car parks, so there is less likelihood of getting a storm water drain with bad air. Please bear in mind, however, that almost all poison gas is heavier than air. In industrial sites, escaped gas can roll along the ground and into the storm water system. Always test the air with the space gas tester for H_2S and oxygen (O^2) content prior to entering a pipeline. It is a good idea to check for hydrocarbons too if the storm water drain is near to a petrol station, fibre glass works, tyre retreader, or refinery.

Where storm water drains are laid over or through reclaimed land, city tips or bail fills, there *will* be gases (methane and others) entering the lines. These lines must be treated as sewers. Never enter any pipe without first testing the atmosphere and then only with outside supervision. Also always ensure that you have the required safety equipment, that it works, and that you know how to use it.

HINTS AND TIPS FOR WORKING IN SEWERS AND DRAINS
Here are some of the golden rules to be considered:

- Never work inside any pipe or manhole alone, no matter what depth. This includes storm water pipes.

- When working in large-bore sewers and drains, fit a length of scaffold pipe tensioned with an acrow prop across the downstream exit pipe hole to assist anyone who may be caught by an unexpected surge current or water flow.

- Whenever possible, work from the downstream entrance up. If anything goes wrong, rescue happens from the downstream exit.

- Always wear a full-body safety harness with a secured lifeline kept gently taut by the observer. Lifeline ends must be tied off to a secure point outside the entry point.

- Surface persons must constantly monitor external conditions that may affect the safety of workers in the hole or up the pipe.

- Wear full-body overalls with sleeves rolled down. Wear a hard hat wherever practical. Wear gloves at all times.

- Do not lean over an open manhole immediately after you open it.

- Inspect all built-in ladders and rungs for corrosion. The most common point of corrosion is at the point where the steel work joins the concrete or brick work. If you were climbing down and a rung gave way, you would fall, breaking one or both legs when you hit the bottom.

- If at all practical, force ventilate all manholes, drains, or pipelines prior to entry. A good flow will rustle your hair as you look into the manhole. Wait at least thirty minutes from the time venting starts before entering.

- Always open the covers on either side of the hole you are entering; it provides an escape route and a place to fit ventilation. In many cases, opening these covers will allow a natural airflow through the pipe. Judicious placement of plywood or canvas can direct wind into the hole and through the pipeline

- Install adequate barriers at least one full metre from and around the open manhole. If you cannot see it from where you are working, fit a secure grille over the open hole as well. One-by-one-inch square egg grating is ideal for this. It needs to be heavy to discourage small children from lifting it. At the same time, it will support the average motor car (in spite of your cones).

- If your men are likely to be down the hole for some time, it is good practice to fit a grating over the working hole as well. A slight trip and the surface supervisor is down the hole on top of the man in the bottom of the hole.

- Never smoke in or near an open manhole or sewer pipe. Keep all flame away.

- Where potential gas risks are likely, carry a space gas tester on you, preferably as close to your head as possible. Very few pipelines are perfectly level (rise and fall), and it is possible for there to be a dip (heavy gas) and or a hump (light gas). Be aware of these.

- Communicate with the man up the pipe; this gives him confidence and lets you know that he is fully conscious and alert. Tell jokes or chat about a football match. You would be surprised at how comforting that is.

- Keep your workmates informed of any illness or disability you may have. Rescuing people often causes harm to others in the process. Never enter a manhole if you are an asthmatic, epileptic, or have a serious hangover. This is no place to have an attack of any kind. You should be one of the surface men.

- Ensure that any person who is giving you medical attention knows that you have sustained an injury working in a sewer.

- Supervise gas tests yourself. It's your life, and you are the man up the pipe. Take the time to learn how the space gas tester works, and use it yourself. Be aware that undisturbed sewerage and decaying matter may not give off gas until you disturb it. The water jet *will* disturb it, and gas *will* be given off.

- Drowsiness or repeated yawning is an indication of bad air. Return to the surface smartly, take a breather, and test the atmosphere with a meter. The oxygen level should read 20.9 per cent. One per cent either way is reasonably acceptable; more than this requires urgent evacuation. A minimum of 17 per cent oxygen is required to sustain life. At 17 per cent, you may not die, but you will be incapable of functioning with any degree of intelligence.

- Any toxic gas, explosive gas, or anomaly measurable on the space gas tester should be a clear signal to vacate the area. It is not unusual for the space gas tester to register small anomalies in the atmosphere. Some of these may not be of sufficient concentration to trigger the alarm. Operators should periodically check the tester (cycle through the screens) to ascertain the condition of the atmosphere. Any abnormal gas or oxygen reading should indicate caution, and ideally workers should carry 10-minute bailout breathing equipment.

- If you are working a retro nozzle up the pipe, it will create a venturi effect and draw back gas that may well rise up the "gas free" manhole and get you. Check, check, and check again. Wear a harness and tie yourself off to your truck. A good burst of gas and you are unconscious and falling into the hole.

- The benching area of the bottom of the manhole, where pipes meet and flow is directed, is slippery and slopes dangerously. Any fall of as little as 150 centimetres into a manhole may break bones. Always enter a new manhole most carefully, and ensure that you are supported firmly by your lifeline or man lifter while descending step irons or ladders.

- If local council regulations require any protection over and above those listed in this document, you must comply with their procedures. If the local council regulations are of a lower standard than described in this manual, you will be guided by the following. Add them; do not subtract for any reason. If you do not have a copy of the local council regulations, ask for them. As an employer, your company is obliged by the HASE Act to identify significant hazards, address them, and inform you of the findings.

- Make sure you have a good communication system between the surface and the observer, and between the observer and the work face operator.

- Work safely. Identify your hazards and address them before starting work. Taking the time to work safely is the smart thing to do.

NOTES ON THE EQUIPMENT AND OTHER PROCEDURES

Breathing Air Compressors
All compressors used for breathing air should have vegetable or non-hydrocarbon-based oils in their crankcase. Most compressors have after-coolers to cool the air; hot air heats the user and makes it very difficult to work. If the air is hot at the face mask, lay the hose in water to try to take some of the heat away.

Know the capacity of the fuel tank, and find out in advance how long the compressor will run on a full tank. Make a clearly visible note of what time more fuel must be added. Get everyone out of the hole and shut down the compressor before refuelling. If anything goes wrong, such as a fire, the compressor will stop. Do not accept shonky compressors for breathing air. If in doubt, don't.

The face mask needs to be supplied with clean, *dry* air to operate correctly. Check the airflow for water droplets. If any are present, fit an extra drier inline.

Always locate the compressor so that the exhaust gas from the engine exits downwind of the compressor's air intake. Watch all vehicles or engines running on site, and make absolutely sure that their emissions are away from the compressor's intake.

Air Filters
Always inform the hirer, if hiring your filters and compressor, that the compressor hired is for breathing air. Also request, in writing if possible, that it and/or the filter pack provided is suitable. If in doubt, make absolutely sure that the filter pack being used has the facility to remove all hydrocarbon material prior to the face mask. Ask the supplier to give you capacities of flow through the filter (how many litres of free air it can filter per minute); this must be able to support more than the number of men breathing from the device. Allow 44 litres of free air per man per minute at a pressure of at least 5 bar (75 psi).

Most filter insert elements have a useful life indicated by a colour change; others work on a time/volume basis. Maintain the filter and be sure you understand how it works. If using hydrocarbon-based compressors (you should *not*), the filter package must contain a section containing activated charcoal or a similar product to remove CO^2 and scrub any oily vapour. Carry spare elements with you. Keep them clean and dry.

Filters should be mounted onto an independent stand. This makes them portable, and you can then carry them off site when you are not working for security. This also facilitates maintenance.

Air Lines for Breathing Air

Surface-supplied air through a 6 mm hose is restricted to a hose length of 60 metres. If you need to go further than that, add in a length of larger bore hose (at least 12 mm) for the extra. Add it at the compressor end and *not* at the mask end.

Garden hose is not acceptable as air line hose. The hose must be rigid or firm, not kink, and be able to withstand being trod on without cutting off the air to the mask. There are a number of specialist hoses manufactured for breathing air lines. Source one of these.

Once a month, wash all breathing hoses by flushing them with a light detergent. Rinse them with a good flow of clean water by attaching them to a freshwater tap for a few minutes. Dry them with clean, dry air. Store with the ends sealed.

Breathing Air Face Masks

Wipe headsets out using a cloth dampened with water and into which has been rubbed a bar of ordinary, plain, unperfumed hand soap like Sunlight Soap. *Do not use dishwashing liquid.* Do not slosh water around into the headset; water will enter into the valves and controls, rendering them suspect at best or useless at worst. Store them in a cotton bag like a pillowcase, and hang them up in a cool place.

Most headsets have the ability to work on demand or free flow. Some units are controlled or switched from one service to another automatically, and others need to be switched manually.

"On demand" means that you need to breathe in to cause the air to flow; "free flow" means that the air runs all the time and you take a breath of it as required. The free-flow method maintains a constant flow/pressure of air through your helmet; this prevents any gas from entering through a poor seal. Some helmets have what is known as a constant pressure regulator, which maintains a pressure of a few psi inside the face mask. As you breathe in, the pressure drops and the bottle tops the pressure back up. This is a perfect system, but it depends on the correct, uninterrupted fit around your face. No beards. If the seal leaks, the unit becomes a free-flow system, and you will use a lot of air.

Breathing Air Bottles

Each person has a greater or lesser lung capacity, and some utilise oxygen better, so there are no hard-and-fast rules regarding useable air. The following gives an indication of expected working times when working with *demand* airflow. For free flow, only 10 per cent of the times below are available.

The small bailout unit has a volume of 2.2 litres, and when pumped up to its maximum pressure of 3,000 psi, it contains 440 litres of free air. This is good for about 10 minutes working hard to escape.

The 6 litre standard rescue bottle contains 1,200 litres of free air and is good for about 30 minutes of hard work when pumped full to 3,000 psi.

The 8 litre bottle contains 1,800 litres of free air and is good for about 45 minutes of hard work when pumped full to 3,000 psi.

All bottles must be tested periodically, and the date of the last test stamped onto the bottle top—normally every five years. All bottles also should have an audible alarm, normally a whistle, which indicates that there is only about 10 minutes of air left.

Be careful who you use to fill your bottles. They must be filled with clean, *dry* air. Any moisture will cause corrosion, and corrosion will render the bottle unsafe—not to mention the taste and the damage moisture can do to the regulator. You have the right to demand that the bottle be filled at a reputable station. Do not use backyard SCUBA club filling stations.

A SCUBA diver died recently when he took his bottles out of his garage after winter, pressure-tested them, found they were full, and went diving. He died in the water. Investigations proved that water in the bottle acting on the steel caused serious rust inside the bottle; this rust consumed the oxygen over the six months they had been in storage.

If a leak develops around the neck of the bottle *do not* try to tighten it while it is under pressure. There is eight tonnes of thrust acting on the base of a valve under full pressure. If you do try, you may break the neck or valve off, and the resulting projectile can travel clean through a half-inch steel plate. Vent it, and get it fixed properly. Never clamp the bottle in a vice.

Maintain a logbook which indicates the next test date, who filled it, and who used it last. If anything goes wrong, you will know where to point the finger.

Do not store full bottles of air for more than six months. Vent them and have them refilled. Air can become stale and taste bad, and any corrosion inside the bottle will destroy the bottle. When venting, upend the bottle, and check for moisture coming out of the valve. Any dampness will require the bottle to be serviced. This involves removing the valve, tumbling the bottle, and doing internal inspection. The person undertaking this work must be qualified to do so. Do not attempt to remove the valve yourself.

Space Gas Testers
There are testers, and there are testers. Some are better or more sensitive than others.

Confined Space Entry

The Draeger bellows unit is very accurate if correctly used. Most contractors have standardised on the Minigas or the Exotox space gas testers, which are constantly on duty and give an audible warning of a change in atmosphere that may become hazardous. Do not shake it to see if it's working properly when it sounds the alarm; it is, so get out now. Test it on the surface, and shake it all you like there.

An earplug adaptor that can be worn under muffs is available. This ensures that you can hear the alarm even in extremely noisy areas.

The unit is not operator repairable; do not attempt to repair it. Calibration needs to be carried out under ideal conditions by a qualified person. Do not fiddle with it.

The unit needs to be charged up and tested in the fresh air. To make sure it works, all you need to do is to breathe out (don't blow) into the sensor grid; it should indicate an oxygen deficiency. It is that accurate. You normally breathe in 20.9 per cent oxygen and breathe out between 16 per cent and 18 per cent oxygen, depending upon your work load. You may have to breathe into it two or three times to make the alarm sound.

Using the Minigas Space Gas Tester

The Minigas tester is one of the more popular, and below is a simple operation instruction for you:

Switch it on by pressing the green button once. Switch it off by pressing and holding the green *AND* red buttons until the screen reads "off".

When you switch it on, it will check itself. It will bleep 12 times rapidly to indicate that its tests were correct. If there is a fault, it will not bleep 12 times, but will announce "fault ###" (that is, the word *fault* followed by a number). You will

High Pressure Water Jetting – An Operator's Manual

need to make a note of this number and send it in with the unit for repair.

The unit will settle in the O_2 scale and should read near 20.9 per cent in the open air. It will keep indicating the oxygen contact on the screen, but it is still monitoring all other functions every three seconds. Should an alarm sound, the screen will indicate which danger exists and will stay in that mode until switched off. Any alarm sounding means you *must vacate the space at once*.

The unit will bleep every 10 seconds to inform you that it is on and functioning. This bleep is quite comforting when you are alone up a pipe.

You can check the other scales by pressing the green button once. This will indicate EXP—explosive gases monitored. It should read 0, or zero.

Press the green button again, and it will read TOX, or toxic gas. This is hydrogen sulphide in most drain units. It also should ead 0 per cent in fresh air.

Press the green button again, and it will read TOXSTEL, which is the Safe Timed Exposure Limits to a toxic gas. These should not concern you. Using the STEL table requires a support booklet and a calculator. If a STEL alarm sounds, vacate the area.

Press the green button again, and it will read TOXTWA. This is toxic gas time weighted average. As for STEL, do not concern yourself with trying to work it out. Just vacate the space.

After a STEL or TWA alarm, it will bleep twice every 10 seconds, meaning that it is reminding you that you are in a STEL or TWA situation and you must plan to record your exposure. You will be out of the space by now, so cancel the alarm. To cancel the double bleep, take the unit out of the vessel or space, switch it off, and wait for 30 seconds before switching it on again.

The unit will hold a charge for about 7 to 10 hours and takes 5 hours to recharge. It is good practice to run the battery flat each day, so leave it on once you switch it on, and put it back into the charger at the end of the day. Switch the unit off before putting it into the charger. If the battery runs flat, the letters bAt will show on the screen. Turn it off and recharged it.

Do not disconnect the battery pack when the unit is switched on. You will damage the printed circuit. If for any reason you wish to change batteries, switch the unit off first.

Testing for Gas

1. Open the manhole, and if at all possible, open one on either side of your hole.

2. Allow a few minutes for the hole to vent.

3. Lower the suction tube into the bottom of the hole, *not* into the water.

4. Pump the bellows at least 12 times.

5. If no alarm sounds, the hole is safe to enter. The person entering the hole still must carry the gas tester into the hole with him.

6. If the alarm sounds, get out of the hole. Ventilate and try again.

It is common practice to lower the tester into the hole, but if there is a high concentration of gas in the hole, it could be great enough to burn out the sensor. If you are using the pipe and bellows, the alarm will sound before the concentration is great enough to affect the element. The cost of the TOX element is

350 dollars, and you could be replacing them about every 4 to 5 months. They should last 18 months. OXY elements are replaced during standard servicing every 6 months; these too cost about 350 dollars, so this is an expensive tool.

Charging the unit correctly is quite important. Do *not* store the unit in its charger during the working day between short periods of use. When you have finished using it, leave it on in the open air, and let it run flat. Ten hours from full to flat is a good yardstick. Then place it into the charger and allow it to charge up again. It takes about five hours from flat to full. Once charged, it may be left in the charger for storage. Store it in a cool, dry place that is well ventilated. It will sound an alarm when the battery is almost flat but will fail to display any signals when completely flat.

All NICAD battery packs develop a memory. If you use 25 per cent of the power and then recharge it, after a while the battery "thinks" its total life span is only 25 per cent and reads flat in that time. Run it all the way down to flat as often as possible to ensure that it doesn't get any silly ideas.

Safety Harnesses

Safety harnesses must be full-body style. Safety belts are not acceptable as fall restraints or for lifting. Ideally they should have a lifting D ring above the shoulders in line with the neck of the wearer.

Safety harnesses must be carefully inspected by a responsible person at least once per month and serviced by a competent person every twelve months. The competent person is most likely to be the supplier. Daily and monthly inspection should include a general condition check and a close inspection of all the strap stitching. More than 12 mm of pulled thread in any *one* line of stitching means that the belt cannot be used and must be sent for repair and recertification.

Confined Space Entry

Check the inside of the webbing at all wear points, such as the places where the D ring rubs, open folds, and inspect the internal surfaces.

Safety harnesses should be washed in warm water with a simple soap, such as Sunlight laundry soap. Rinse them carefully, and hang them up to dry. The units are UV protected, so no harm will be done to them if they are hung in the sun. *Do not use petrol, kerosene, solvents, industrial degreasers, or hand cleaners* to clean safety harnesses.

All safety harnesses are marked with a serial number by the manufacturer. These numbers should not be defaced and should be entered into a safety-equipment logbook to enable you to gather historical data regarding the life and testing of the belt.

To put on a safety harness, proceed as follows:

- Loosen all the buckles.

- Place the diamond-shaped patch (serial number facing up) in the palm of your hand with the D ring over your fingers and the leg straps on either side of your forearm. Shake out all the tangles and twists in the straps.

- Take hold of the D ring, lift the unit over your back, and place your arms through the shoulder straps one at a time.

- Tighten the shoulder straps until the waist belt is about level with your trouser belt line.

- Thread the waist belt through the belt buckle, and secure it loosely on the Velcro patch.

- o Pass one leg strap through your crotch from the front to the back, and clip it on. Repeat for the other leg. Do not tighten the leg straps yet.

- o Standing up straight, snug up the waist belt with sufficient slack for you to place your hand sideways between your stomach and the belt (about 100 mm). CAUTION: This slack is most important. Should you fall with the belt too tight, the belt buckle could jerk up under your rib cage and rupture your spleen.

- o Squat and tighten the leg straps snugly. Stand up straight, and you should be quite comfortable.

- o Tuck all excess straps under your waist belt to prevent them flapping about.

Lifelines or safety ropes

These lines are normally 12 mm nylon ropes which have eyes spliced into each end. They are life-support equipment and are not for towing the car, for using as a crane sling, or for sitting on. They are needed to save your life. Keep them clean and dry, and in a cloth bag out of the sunlight.

All harnesses are fitted with a 1.8 metre tether or short lifeline. This remains permanently on the harness. This length of line is the furthest you should ever be able to fall. More than 1.8 metres, and you will do some serious and permanent damage to yourself. This is your "tie myself off" line; one end is attached to the D ring around your back, and the other is used to clip onto whatever support there is around. Your long lifeline is also attached to the end of this line. The man lifter or fall arrester goes directly onto your D ring.

These safety ropes or lifelines must be inspected carefully at least every thirty days. At wear points, the rope should be carefully unlaid to allow you to inspect the rope in the centre of the lay. Any sign of little bits of fluff or wadding lying loose amongst the inner braids is an indication that the rope is beginning to decay. It's time to replace it. Any external damage that frees one thread of one strand is also an indication of the rope's pending failure. Cut of the eyes from each end, and use it as a work rope for tying things up.

If ever you need to wash your rope, take it home, lay it in the bath, add some warm water, and a little Sunlight laundry soup or Lux flakes, and walk on it barefoot, like pressing grapes. Rinse it carefully and lay it over a clothesline in zigzags.

To get the twists and kinks out of a lifeline, tie one end to a strong point, pull it tight and straight, and "milk" it, somewhat like milking a cow. All kinks and twists will come out this way. Coil it up using a figure-eight formation, and fold the two O's together to turn it into a single circle. It will unlay it without any twists in it when you need.

MAN LIFTER, FALL ARRESTER, AND "A" FRAME
The man lifter kit normally includes the following:

1. An A frame with adjustable legs, a circumferential leg restrainer, and normally a short chain with an eyelet in the top for hanging the man lifter/fall arrester gear.

2. A carry bag containing the fall of rope, normally 12 mm diameter, braided nylon line. This rope is of sufficient length to provide the required lift height multiplied by four (four falls of rope).

3. A man lifting block and tackle. This incorporates the lifting pulleys, fall arresting clutches, and swivels.

The unit needs to be carefully looked after. There is little or no maintenance other than a daily visual check, but it *must* be kept clean. Cleaning the block and tackle is best accomplished using an air line fitted with an airgun to blow away dust and dirt. Do not pull it apart; this should be done by the supplier. The rope should not need cleaning if it is properly looked after, but if it does, wash it in warm water using a mild soap like Sunlight laundry soap. Carefully rinse it and stretch it out to dry. Never use any form of solvent on the rope.

Setting Up the Man Lifter

- Take the unit from its bag, and lay it on the floor with the top of the A frame away from the hole.

- Open two legs (one on either side), and extend them to the required height.

- Lift the unit up and over the hole with the third leg being carried across the hole. The legs should be positioned so that the lifting eye is directly over the hole. If the ground is sloping, adjust one or more legs longer than the other(s) to ensure that the line pull is straight from the lifting eye. The ideal height leaves enough room for the man being lifted to sit on the edge of the hole plus the height of the almost closed block and tackle.

- Once set up, fit the circumferential chain to the base of each leg to stop the legs moving. In some cases, it may be practical to peg the feet of each leg through the hole provided with a length of reinforcing steel driven into the ground.

- Stretch out the rope fall, removing all twists. This is best done by tying one end of the rope to a strong point, pulling the rope out straight, and "milking" the end (a motion not unlike milking a cow). This will cause all the twists to run towards you and out of the end of the rope.

- Rehang the block and tackle, and figure-eight the rope carefully on the ground near where you are to work.

Going Down

- Check the safety harness to ensure that it is correctly fitted, and sit the man in the hole. He sits on the surface with his feet in the hole.

- Attach the karabiner to the man's safety harness, and upend it after doing up the safety thimble. Safety thimbles have a coarse thread and are screwed up over the opening part of the karabiner. It is advisable to turn the karabiner so that the thimble hangs down (towards the tightened position). This way gravity cannot cause the thimble to unscrew itself.

- Take the strain on the block and allow the man to enter the hole. Pay out the line gradually as he goes, and allow his weight to slowly pull the line though the block.

Should he slip or fall, the fall arrester will lock up within 100 mm (4"), and he will be comfortably supported by his harness. To unlock the fall arrester, pull sharply on the line; this will take a pull of about 50 mm to unlock the fall arresting latch. Once settled and ready to proceed allow the man to proceed once again.

Coming Back Up

There are three ways to get a man back up: surface men lift him (done easily by one man with ideally another tending the lines); he lifts himself (he needs to have the rope tail in the hole with him); or you help him out while he climbs a ladder or step irons (you take up the slack as he comes).

Rope Twisting

This is a common occurrence which happens because you did not get all the twists out of the line before you started. The problem is not a serious one, as the rope will unwind as the man gets close to the surface. It does, however, cause a great deal of friction as each rope rubs against the other, causing the top man to use a great deal of effort to lift a simple load.

Packing Up

Take all the slack out of the block and tackle (pull the blocks almost together), and coil your rope in a Fig 8 fashion, lay the blocks inside the hole of the 8. Using a shoelace or a piece of twine, tie the rope together at a point in the bottom and top of the 8, fold the top towards the bottom, and place the 8 (now an O) into its carry bag. Using it next time will be no trouble.

PERMITS TO WORK
A permit to enter a confined space is common in industrial applications, but few councils provide them for entry into their sewers or drains. This is a shame, as there is much information that can be shared. These dictate the owner's safety requirements, identify significant hazards, and state minimum safety requirements.

Permits to work should be required in the following circumstances:-

- Manholes and sewers known to contain gas

- Manhole and chambers deeper than 7.0 m (22'6")

- Unventilated culverts or culverts with poor access

- Workplaces known to be hazardous

- Trunk sewers 1.5 m (5') diameter and over

If you are issued with a Permit to Work, it is your duty to comply with the conditions laid down on it. It is essential that you are properly trained in the use of permits.

RAT POISON
Extremely toxic rat poisons are used in some manholes. The rat bait is normally grain coated with the poison and is left on manhole benching, normally in small containers or plastic tubes which the rat can enter. The poisoned grain is bright yellow in colour when fresh, but this degenerates into a brown, translucent jelly.

Protective clothing should be worn when entering manholes, and a watch should be kept for rat bait containers of any kind. On no account should rat bait be touched. If protective clothing is accidentally contaminated, wash it down thoroughly before removal, and do not touch the contaminated parts with exposed skin.

If rat bait is seen and work cannot be completed without risk of touching the bait, the manhole should first be water jetted clean before entering. If rat bait is removed, the pipe owner shall be informed. Wear appropriate protective clothing. Do not touch dead rats; remove them with a shovel and bucket.

ELEVATED MANHOLES
When entering elevated manholes, suitable and safe access to the cover must be provided, so far as is reasonable and practicable. Generally, this will apply to manholes protruding above ground level by more than 1 metre and will take the form of scaffolding erected in accordance with the standard.

Elevated Sewers and Drains

When a sewer or drain is at such a level above ground as to make it unreasonable to work on the sewer while standing at ground level, scaffolding must be used. Ladders may be used up to a height of 1.98 metres for access to a sewer or drain. If this height is more than 1.98 metres [6.6'] then scaffolding should be used. If this is not practicable, then slung or suspended scaffolds, EWPs, bosun's chairs, or similar should be used.

In all the above cases, the hazards that may be encountered should be considered most carefully, and all practical steps must be taken to minimise the risk to employees.

WORK ON WATER

Persons engaged in work in deep or dangerous waters or where there is a danger of falling into deep water should be able to swim and must wear life jackets. When a person is working from a boat in a fast-flowing river or near weirs, a tethering line attached to the boat and another line attached to the worker must be secured to the bank or held by a shore party to prevent the boat or worker from being swept downstream.

The use of makeshift rafts is not permitted.

OCCUPATIONAL SAFETY AND HEALTH INSPECTORS

Health and Safety inspectors have a legal right to enter into or to visit any site premises at any reasonable time. If you are approached by such an inspector, you must follow the procedure set out below.

1. Ask him to produce an official means of identification (News reporters have been known to pretend to be inspectors when you are working on a sensitive environmental site.)

2. When you are satisfied with his identification, give him all the help he may ask for (if you obstruct the inspector, you could be prosecuted.)

3. If an inspector appears to be making requests which in your opinion could endanger property, personnel, or third parties, refer the inspector to your immediate supervisor.

IMPROVEMENT OR PROHIBITION NOTICES

1. If the inspector deems it necessary to issue either an Improvement or a Prohibition Notice to you, you must accept it without argument.

2. When the notice is in your possession make every effort to notify your supervisor without delay.

3. Senior management and the safety officer must also be notified and the notice handed to management in a clean condition.

All employees should cooperate with OSH safety inspectors. It is particularly desirable that supervisors build up a good working relationship, which can best be secured by appropriate full communication with the safety inspector and joint cooperative effort in workplace inspection and accident investigation. Any changes in working practices should be discussed with the safety inspector who should be kept generally informed and particularly informed when Permit to Work situations arise.

TO SUM IT ALL UP

Finally, enter all manholes with care. Do not get casual or blasé about your work; just when you think you've got it all sussed out, it will turn around and bite you. Do not take chances and think carefully.

Play the "what if" game regularly: one man asks, "What would we do if so and so happened?" The others work out the action to be taken. This should be done during meal breaks and smoko. Make a plan and adopt it as procedure. If you are the man in trouble, you will already know what the others will be doing to get you out of trouble.

Identify all your hazards. Eliminate, isolate, or minimise all the significant hazards, and take time to consider safer ways of doing the job. What we do is dangerous; make it safer.

MINIMUM SAFETY EQUIPMENT AND TRAINING REQUIREMENTS

The following specific instructions must be read and fully understood. The point is, if in doubt *don't*.

Personal Protective Clothing and Equipment
For all employees on site:

- Safety boots or shoes
- Hard hats
- Safety eyewear appropriate for activity
- Gloves
- Overalls with long sleeves
- Rain wear
- Health and hygiene facilities:
- Rubberised gloves

Confined Space Entry

- First-aid kit (complete) and accident register
- Disinfecting hand-washing equipment
- Drinking water
- Reporting and help facilities
- Communications device, mobile phone, or radio
- Accident reporting forms
- Equipment logbook
- Entry checklist
- Notebook
- At least one pair of hand-held radios
- Bottom or entry persons (this includes equipment for the standby rescue person on the surface, where appropriate):
- Space gas tester (individual if practicable)
- Rubber boots or waders
- Rubberised work gloves
- Wet weather wear, oilskins
- Bump cap (if hard hat not practical)
- Individual full-body, waistcoat-style safety harness (rear D ring)

- 1.8 metre lifeline tether fitted with nylon eyes and fail safe karabiners

- Extra 12 mm lifeline as needed for depth or length of ingress, one end spliced with an inserted eye

- Intrinsically safe lighting

- Flotation vest for deep water applications

- Walking staff, if needed

Safety Equipment Needed, Depending on Classification

- Standards-approved man lift and fall arrester with fall line suitable for the depth

- Standards-approved A frame

- Spare lifelines

- Rescue BA set

- Bailout BA sets

- Full BA sets

- Complete set of road signs as needed for the site if work is in roads

Training and Special Skills
All crew members "should have the following:

- Confined space training

- Emergency procedures

- Evacuation procedures

- BA trained (fit and use)

- First Aid and CPR (an ideal situation, should be more than one)

- Space gas detection (responses)

- Man winch use

- Working on the roads, traffic signs and their placement

Responsible Person (Job leader)
The person must have all of the above, plus the following:

- First Aid and CPR current and up to date

- BA trained (fit, use, care, and maintenance)

- Safety supervision and hazards identification procedures

- Permit and entry procedures

- Communication and reporting procedures

- Space gas testing (master set) equipment and procedures

- Man winch use, care, and maintenance

- Safety equipment inspection procedures

NOTE: The responsible person should be authorised in writing by his employer and by the client as responsible for the safe operation of the project. The responsible person's position is one of safety. He is not an engineer or professional person and needs no special skills or abilities. He is, however, ultimately responsible for the safe conclusion of the project. His responsibilities override all others.

Notifications to Be Given

To the owner of the system:

- That you are about to enter
- Number of persons entering
- Poor structural integrity
- Urgent repairs or inspections needed
- All workers out of the system upon completion

To the local fire station:

- Location of the job, from start time—to finish, depth and pipe size, if known
- Number of persons entering
- Approximate completion time
- Job complete, thank-you

Occupational Safety and Health inspectors (twenty-four hours in advance):

- If BA is being used
- If significant hazards exist

Local council (twenty-four hours in advance):

- If you intend to close lanes on a major road way

JOB CLASSIFICATION AND MANNING LEVELS

Job-site inspection

Primarily you must always consider doing the job from the surface. Is it really necessary to put a man in the hole or up the pipe? If it can be done from the surface, it is well worth any small cost that may be incurred making special tools to do it. Any risk or hazard by which workers are exposed to serious harm should be eliminated if at all possible. Entering a manhole is dangerous enough. Going up a pipe is playing Russian roulette with someone's life.

Under ideal conditions, all jobs should be inspected a few days before they are due to start. I understand that this is not always possible, but you should try. The following information is needed:

1. A ground plan clearly showing the location of manholes and the destination of the pipe.

2. The road layout if the job is in the roadway. A sketch should be made to identify traffic control signs that will be needed. Identify peak traffic periods, and work outside those times if possible.

3. Manhole depth if the client does not know, which is the norm. Open the cover and drop a tape in—*after* fitting and securing your harness.

4. Pipeline diameters and estimated flow (quarter, half or full). Can you do the job with this amount of flow?

5. Function of the pipe, sewer, or storm water; estimate peak flow periods and work outside these times.

6. Expected weather conditions at time of job, flash floods, and full pipes if drain or combined sewer and drain (fairly common in Auckland).

7. Lower the space gas tester and check for gas present. Consider ventilation procedures if gas is indicated.

This inspection should be carried out by the responsible person; he should have the necessary skills to identify the hazards of the job and to be able to classify and select manning levels and equipment levels needed.

In the event of a call out, the operators sent to the job should classify the job and identify the hazards. If these are greater than anticipated for the tools and manpower available, the job should not be undertaken. Call your base controller to inform him of your needs. Do not take a chance. Do it right first time, and you will live to talk about it.

CLASSIFY THE JOB

NOTE: The indicated manning levels are for manhole entry only. Extra men are required if pipeline is to be entered.

Class Three

This requires 3 men (or 2 with man lifter), 1 x safety harness and lifeline, and a space gas tester,

- Storm water manholes less than 3 metres deep
- Sewer manholes less than 2.5 meters deep
- Pipelines 300 mm or smaller
- No gas indicated or expected
- No sudden flows expected

Confined Space Entry

Class Two
This requires 3 men and a man lifter, a first aid and CPR-qualified top man, 2 x safety harness and tether, a space gas tester, a bailout BA, and a rescue BA to be on site and utilised if any gas is suspected or indicated during job. In the event of serious gas indications, the job should be stopped and reclassified as a Class 1 Gas Entry.

- Storm water manholes less than 6 metres deep
- Sewer manholes less than 5 metres deep
- Pipe sizes bigger than 300 mm
- No gas indicated
- No sudden flows expected
- Ventilation possible (mechanical or natural)

Class One
This requires the responsible person plus three men, a man lifter, 2 x harness and tether, a space gas tester, a bailout BA, and a rescue BA if gas is possible. If serious gas is indicated, the job should be stopped and reclassified as Class 1 Gas Entry.

- Drains more than 6 metres deep
- Sewers more than 5 metres deep
- High flow rates
- Possibility of sudden flows
- Possibility of gas present
- High risk of trade waste contamination

- Dangerous or difficult access

- Extreme levels of known contaminants

- Uncontrollable atmosphere

MANNING LEVELS

All confined space entry should be supervised by a responsible person. For all Class 1 work, a responsible person who is fully trained and conversant with all the procedures is essential. The responsible person does not enter the pipe unless he has a qualified standby man to replace him. The standby man should have similar qualifications as the responsible person.

Manhole Entry

- Three men for inspection only

- Minimum of two surface men—or one surface man if man lifting and fall arrester device is used

Pipeline Entry

- Four men (pipe entry for local repair)

- Two on surface or one if man lifter is used, one observer at bottom of the manhole, one (or more if needed) man working in the pipe

Pipeline Inspection

- Six men (walking the line between manholes)

- One man on surface at each manhole, one at bottom of each manhole and two walking the pipe no more than 6 metres apart

Confined Space Entry

Entry into Class 1 Gas

- Five men (for manhole entry)

- Two on the surface tending the man lifter and entry lines, one standby rescue man, one responsible person, and one in the hole. For pipe entry (gas), this number will increase.

Vessel or Space Entry

Class 3 Entry (no pipe entry required)

Prior to all entry, the responsible person must personally and visually check all safety gear to be used by his crew (see Entry checklist at rear of this book). This will include but not be limited to the following checks and actions that need to be addressed:

- Client informed of entry

- OSH informed, if notifiable job

- Fire rescue people informed (if major significant hazards exist)

- Toolbox meeting completed, significant hazards identified and addressed

- All crew aware of the procedures you intend to use in the event of an emergency

- Space gas tester fully charged and calibrated (note service date; is it in date?)

- Harnesses: check all stitching and webbing, buckles, and D rings

- Rescue BA (if sewer), laid out on tarp, buckles adjusted, bottle full, valve on, and face masks tested

High Pressure Water Jetting – An Operator's Manual

- o First-aid kit checked and complete

- o Lifelines: no damaged ply, no frayed ends, untangled, and identified by name

- o Flashlights: safe to use, batteries in good order

- o Check all crew for open wounds—disinfect and dress any with waterproof dressings

- o Rubber or waterproof gloves for all bottom men inspected to ensure no holes

- o Personal protective clothing, as required

- o A frame secured, and if needed pegs driven, block line fully extended and untwisted, laid in large figure-eight on the ground

- o Strong point for lifelines available

- o Communication devices work, test the mobile phone for S service strength, battery fully charged

- o Alternatively check the two-way radio to make sure it works and has a good signal

If you cannot answer *yes* to all of the above, do not enter the space.

Entry:

This is the basic inspection procedure where a man enters a shallow manhole to rod, mirror, or push a video camera up or down the pipeline. No entry into the pipeline is intended. The pipe is small bore, and the man could not be washed down it in the event of a sudden flow.

Confined Space Entry

- Three men (or two if a man lifter is used).

- Where possible, manholes on either side of the access point will be lifted to provide escape and ventilation.

- Ventilation should be carried out for 30 minutes prior to entry. This is an essential practice when working in sewers or leachate manholes

- The space gas tester will be used according to the manufacturer's instructions to ensure no gas is present.

- If gas or oxygen deficiency is indicated, the job classification must change to a Class 2 entry.

- Establish emergency evacuation procedures; include all team members in this.

- The man entering the hole will carefully inspect the integrity of the structure prior to entry.

- The man entering the manhole remains secured to the man lifter or his lifeline at all times. The end of the lifeline will be secured to a strong point on the surface.

- The top man will maintain constant tension on the lifeline until the bottom man is on the bottom.

- The top man will provide only enough slack for the bottom man to do his job and then tie off the lifeline to a strong point.

- The top man will maintain communication with the bottom man throughout the period of entry.

- Upon completion of the job, the top man will take up the slack on the lifeline and assist the bottom man out of the hole.

Class 2 Entry (no pipe entry expected)

- Two men on the surface plus the man lifter (or three men), one man down the hole

- One surface man will be dressed in a full-body safety harness with lifeline attached, ready to carry out an injury rescue if required.

- Where possible, manholes on either side of the access point will be lifted to provide escape and ventilation.

- Ventilation should be carried out for 30 minutes prior to entry. This is an essential practice when working in sewers or leachate manholes.

- Establish emergency evacuation procedures, and include all team members in this.

- The space gas tester will be used according to the manufacturer's instructions to ensure no gas is present.

- If the possibility of gas is expected (all sewers) but not indicated on initial testing, the bottom man will carry the space gas tester with him and will also wear an escape BA set ready for use.

- If gas is indicated during initial testing, the classification must change to a Class 1 gas entry, or mechanical means of ventilation are obtained and ventilation continues until the atmosphere is safe to enter without full BA.

- The man entering the hole will carefully inspect the integrity of the structure prior to entry.

- The man entering the manhole remains secured to the man lifter or his lifeline at all times. The end of the lifeline will be secured to a strong point on the surface.

Confined Space Entry

- The top men will maintain constant tension on the lifeline until the bottom man is on the bottom,

- The top men will provide only enough slack for the bottom man to do his job and then tie off the lifeline to a strong point.

- The top men will maintain communication with the bottom man throughout the period of entry.

- Upon completion of the job, the top men will take up the slack on the lifeline and assist the bottom man out of the hole.

Class 1 Entry (no pipe entry expected and no gas present)

- Four men, one of whom is a trained responsible person, plus a man lifter.

- The responsible person directs and controls entry and exit.

- One trained surface man will be dressed in a full-body safety harness with lifeline attached, ready to carry out an injury rescue if required.

- Where possible, manholes on either side of the access point will be lifted to provide escape and ventilation.

- Ventilation should be carried out for 30 minutes prior to entry. This is an essential practice when working in sewers or leachate manholes.

- The space gas tester will be used according to the manufacturer's instructions to ensure no gas is present.

- If the possibility of gas is expected (all sewers) but not indicated on initial testing, the bottom man will carry the space gas tester with him.

- If gas is indicated during initial testing, the classification must change to a gas contaminated space entry, or mechanical means of ventilation are obtained and ventilation continues until the atmosphere is safe to enter without full BA.

- Establish emergency evacuation procedures, and include all team members in this.

- The man entering the hole will carefully inspect the integrity of the structure prior to entry.

- The man entering the hole remains secured to the man lifter or his lifeline at all times. The end of the lifeline will be secured to a strong point on the surface.

- The top men will maintain constant tension on the lifeline until the bottom man is on the bottom,

- The top men will provide only enough slack for the bottom man to do his job and then tie off the lifeline to a strong point.

- The top men will maintain communication with the bottom man throughout the period of entry.

- Upon completion of the job, the top men will take up the slack on the lifeline and assist the bottom man out of the hole.

Entry for work inside a pipe (No gas)

- Minimum of three surface men, one of whom shall be the responsible person, plus one observer in the bottom of the manhole and one or more working inside the pipe. There must always be enough men or equipment on the surface to pull an injured man out of the hole. The number of persons working inside the pipe should be restricted to the bare

Confined Space Entry

minimum. In the event of an urgent evacuation, there is only one ladder or lifter to get the men out.

- One surface man will be dressed in a full-body safety harness with lifeline attached, ready to carry out an injury rescue if required.

- Where possible, manholes on either side of the access point will be lifted to provide escape and ventilation.

- Ventilation should be carried out for 30 minutes prior to entry. This is an essential practice when working in sewers or leachate manholes.

- Establish emergency evacuation procedures, and include all team members in this.

- The space gas tester will be used according to the manufacturer's instructions to ensure no gas is present.

- If the possibility of gas is expected (all sewers) but not indicated on initial testing, the man up the pipe will carry the space gas tester with him and will wear an escape BA set ready for use.

- The bottom man (observer) entering the hole first will carefully inspect the integrity of the structure prior to entry.

- The man entering the manhole will remain secured to the man lifter or his lifeline at all times. The end of the lifeline will be secured to a strong point on the surface.

- The top men will maintain constant tension on the lifeline until the bottom man is on the bottom.

- The top men will provide only enough slack for the bottom man to do his job and then tie off the lifeline to a strong point. If the man lifter is being used, the bottom man (observer) frees himself from the lifter while remaining tethered to the surface with his lifeline. It is essential to eliminate all slack to prevent confusion and tangles below. The tethers must be clearly identified on the surface.

- The bottom man (observer) will establish a safety pipe across the entrance to the downstream pipe entrance.

- The man or men entering the pipeline will now follow; each will have a lifeline and be attached to the man lifter (if used) during his descent.

- The bottom man or observer may duck into the pipe while this takes place.

- Once the pipeline workers are on the bottom, the bottom man (observer) refits the man lifter (if used) to his harness. In the event of an emergency, he is removed first, once the pipeline worker is on his way back to the exit point.

- The worker in the pipe will have a lifeline, with the end secured on the surface, which will be monitored and slack controlled by the bottom man or observer.

- If more than one man is working in the pipe, they may be joined by a tether no more than 6 metres long. Better practice is for each to have his own lifeline which is colour or size identifiable. It is essential to keep the crew in the pipe as small as possible. In a majority of cases, the work can be done by one or sometimes two. More are a serious hazard.

- The bottom man will maintain communication with the pipeline worker(s) at all times. This communication should be visual, but if a bend exists, the workers must talk to each other to ensure that they are alert and not in any difficulty. It is good practice to talk or joke to bolster the confidence of the man up the pipe. Imagination can play funny tricks while confined inside a pipe deep underground.

- The top men will maintain communication with the bottom man throughout the period of entry.

- Once work commences and the pipeline man is on site, all lifeline slack should be removed from the hole and tied off on the surface

- Upon completion of the job, the top men will take up the slack on the lifeline and assist the bottom man (observer) out of the hole.

- The pipeline worker or workers will follow in order.

Class 1, Pipe entry for line inspection (walking the pipe) no gas present)

- Minimum of six men plus the responsible person, one man lifter set, one top man at each hole with radio or visual communication with each other, one observer in the bottom of each manhole and two walking the pipe. There must always be enough men or equipment on the surface to pull an injured man out of the hole.

- One surface man will be dressed in a full-body safety harness with lifeline attached, ready to carry out an injury rescue if required.

- Where possible, manholes either side of the access point will be lifted to provide escape and ventilation.

- Ventilation should be carried out for 30 minutes prior to entry. This is an essential practice when working in sewers or leachate manholes.

- Establish emergency evacuation procedures, and include all team members in this.

- The space gas tester will be used according to the manufacturer's instructions to ensure no gas is present.

- If the possibility of gas is expected (all sewers) but not indicated on initial testing, the men walking the pipe will carry the space gas tester with them and will also wear an escape BA set ready for use.

- One surface man will be dressed in a full-body safety harness with lifeline attached ready to carry out an injury rescue if required.

- The space gas tester will be used according to the manufacturer's instructions to ensure no gas is present.

- The bottom man (observer) entering the hole first will carefully inspect the integrity of the structure prior to entry.

- The man entering the manhole remains secured to his lifeline at all times. The end of the lifeline will be secured to a strong point on the surface.

- The top men will maintain constant tension on the lifeline until the bottom man is on the bottom.

- The top men will provide only enough slack for the bottom man to do his job and then tie off the lifeline to a strong point. If the man lifter is being used, the bottom man (observer) frees himself from the lifter while remaining tethered to the surface with his lifeline. It is essential to eliminate all slack to prevent confusion and tangles below. The tethers must be clearly identified on the surface.

Confined Space Entry

- The bottom man (observer) will establish a safety pipe across the entrance to the downstream pipe entrance.

- The men entering the pipeline will now follow; each will have a lifeline and be attached to the man lifter (if used) during his descent.

- The bottom man or observer may duck into the pipe while this takes place.

- Once the pipeline workers are on the bottom, the bottom man (observer) refits the man lifter (if used) to his harness. In the event of an emergency, he is removed only when the pipeline walkers are well on their way to the upstream exit point and are in contact with the observer in the next manhole.

- The men walking the line should be joined by a tether no more than 6 meters long.

- The bottom man will maintain communication with the pipeline walkers until they are in contact with the observer at the exit point. This communication should be visual, but if a bend exists, the workers must talk to each other to ensure that they are alert and not in any difficulty. It is good practice to talk or joke to bolster the confidence of the man up the pipe. Imagination can play funny tricks while confined inside a pipe deep under the ground.

- The top men will maintain communication with the bottom men throughout the period of entry.

- Once the walkers reach the exit point, are secured, and are ready to leave the pipe, the downstream observer will remove the safety pipe and leave for the surface.

- The top man and responsible person now proceed to the exit point with the man lifter, if used, to assist in the removal of the crew.

- Safety lines will be lowered to ensure safe exit while climbing out of the exit point. *No one may climb out of or into a hole unless secured to a lifeline that is kept taught and supported by others.*

Class 1, manhole or pipeline entry with gas present

At this point we should stop and seriously consider what we are about to do. Any gas-contaminated or oxygen-deficient space is a serious hazard area. People can get killed. Is there no way you can provide ventilation? There are big blowers around available for hire as well as air syphons or movers available for hire. The cost of the ventilation is nothing compared to the potential loss of life of men entering contaminated spaces.

Set up ventilation so that it blows into the hole; only sucking gas out is not a good idea. You could set one blower in one hole and a sucker in the other to cause air circulation. You must provide breathable air in there. Also set up your ventilation in a way that it does not restrict entry or emergency exit.

We will assume that you have investigated all the alternatives in depth. Remember the Health and Safety in Employment Act? We must eliminate, isolate, or minimise all hazards. Have you really tried? Inform OSH of your intentions twenty-four hours prior to starting the job.

Space entry

- Responsible person, standby person, two top men (one a qualified first aider), man lifter, bottom man—the observer if a pipe entry and the man up the pipe.

Confined Space Entry

- Three sets of breathing gear are needed, two if it's a manhole entry only. All electrical equipment used in the space must be classed "intrinsically safe". The man or men entering the space should preferably have surface supplied air line breathing apparatus with 10-minute bailout bottles attached to their harness. Try getting into a confined space with a bottle on your back.

- Operators wearing bottled air must be monitored at all times to ensure that they do not overstay their air supply. Each operator must have a well-fitting, correctly adjusted, positive-pressure headset that has been tested and adjusted *prior* to entering the manhole. Beards are to be removed; a good seal cannot be obtained over a beard.

- If surface-supplied air is used, the air supply should be quad (bottled breathing air available in four or six bottle quads) with adequate spare air. Alternatively, use a compressor with a reservoir of at least 10 minutes of air for all the workers being supplied. The compressor capacity should be at least 44 litres of free air per man per minute and operate at a pressure of at least 75 psi at that flow. Should the compressor fail or break down for any reason, you must ensure that there is enough air in the reservoir to get the men out of the line before it runs out: 44 litres x men x minutes. Remember that as the pressure drops, so does the available volume, so the pressure available in the reservoir must be able to be maintained at or above 75 psi to still keep the face masks working. The surface operator must ensure that he starts the men out as soon as the compressor fails. He should not try to restart the unit until he has the men moving out.

- If the compressor does not have a reserve supply, the operators must carry 10-minute bailout bottles of air attached to their breathing apparatus. A good communication system must be established. All operators must be trained in the use of the bailout facility. Practice this activity one day prior to entering the confined space; let them breathe on bailout supply for several minutes. Shut off the compressor supply so that they are aware of the sensation and know how to cross over from one source to another, and then recharge the bottles.

- The standby or rescue person must be dressed with a harness and breathing apparatus fitted. All he needs to do is put on his face mask and proceed with the rescue

- Establish emergency evacuation procedures, and include all team members in this.

- A method of communication using rope tugs should be established to pass messages, as it is difficult to understand someone talking in a full face mask

- All entry will be logged by time and name. It is essential that the responsible person is aware of the volume of air in a bottle and how long that will last, and ready to remove the man when he has 10 minutes of air left.

- The standby man should have an alternate breathing air supply to that being used by the workers in the space. This is because, should there be a need for rescue, it would most likely be because of a failure in the air supply to the men in the space.

- The bottom man (observer) entering the hole first will carefully inspect the integrity of the structure prior to entry.

- The man entering the manhole remains secured to the man lifter or his lifeline at all times. The end of the lifeline will be secured to a strong point on the surface

- The top men will maintain constant tension on the lifeline until the bottom man is on the bottom

- The top men will provide only enough slack for the bottom man to do his job and then tie off the lifeline to a strong point. If the man lifter is being used for other man entry, the bottom man (observer) frees himself from the lifter while remaining tethered to the surface with his lifeline. It is essential to eliminate all slack to prevent confusion and tangles below. The tethers must be clearly identified on the surface

- The bottom man (observer) will establish a safety pipe across the entrance to the downstream pipe entrance if men are to enter the pipeline.

- The men entering the pipeline will now follow; each will have a lifeline. It is good practice to tape the air supply hose to the lifeline, which eliminates excess tangle and things to take care of. The men must be attached to the man lifter (if used) during descent.

- The bottom man or observer may duck into the pipe while this takes place.

- Once the pipeline workers are on the bottom, the bottom man (observer) refits the man lifter (if used) to his harness. In the event of an emergency, he is removed first, but only when the pipeline walkers are well on their way back to the exit point.

- The top men must bring in the hose/lifeline umbilical as the workers come out. A light tension should be kept to indicate movement.

- The top men will maintain communication with the bottom men throughout the period of entry.

- The last worker out will remove the safety pipe before he leaves for the surface.

- *No one may climb out of or into a hole unless secured to a lifeline that is kept taught and supported by others or a strong point.*

Culvert entry, no gas present, 3 men

These instructions apply to culverted watercourses with no indication of pollution, gas, or oxygen shortage and with good access and egress from both open ends. If these conditions do not apply, the culvert will be classified as Class 1 or 2 Confined Spaces, and all requirements for the classification will apply. In deciding manning levels, account must be taken of the work location in relation to the availability of assistance in emergencies.

- Locate nearest telephone or, if available, radio telephone.

- Do not enter unventilated culverts or culverts with poor access without informing the owner and rescue services prior to entry.

- Remember that dry culverts can contain gas or be deficient in oxygen.

- Test for gas and oxygen deficiency. If detected, keep out and reclassify the entry.

- Before entering the culvert, remove access covers and ensure that rapid egress from the culvert can be gained in an emergency. Check steps irons, ladders, and so on, at points of egress, and place ladders if necessary.

- Carefully inspect the structural integrity of the culvert prior to and as you enter. Any indication that the culvert is not structurally sound should indicate an immediate exit and a report back to the owner.

- Culverts which can be walked are to be walked by a minimum of 2 men keeping within range of top man's voice call. If practical, the two men should be roped together by a lifeline no more than 6 metres long. A gas and oxygen space gas detector, switched on, must be carried by one of the men walking the culvert.

- If the water is deep or the culvert bottom is rough, a lifeline to the top man should be worn as well as a flotation collar or life jacket. If necessary, use a staff for support to feel the way ahead.

- The top man is to contact men walking culvert at frequent intervals to confirm that they are in good spirits and not in any difficulty. The top man must be aware of what is happening around the area of the culvert and to give notice of pending rainfall.

In the event of no reply to a top man's call, the top man should take emergency action—that is, telephone or radio for immediate assistance from the 111 fire rescue services.

Do not enter any culvert where

- structural stability is in doubt,

- height is less than 1.0 m (3.3"),

- width is less than 600 mm (2'0"), or

- access points are spaced at intervals which do not allow good communication between top man and persons walking the culvert.

Emergency Procedures and Responding to Emergencies

What is an emergency?

- Any incident, happening, sequence of events, action, or reaction that may endanger the crew or any individual of that crew.

The identification of an emergency before it becomes one is the mark of a good responsible person or job leader.

Responding to an Emergency

If all emergencies (hazards) are identified and their consequences planned for, any emergency situation will be well managed, orderly, and without risk. Rescuing people in a panic situation often injures more people than the injury initially caused. All eventualities must be identified, discussed, and acted on; the reaction is planned in detail. This is one of the prime roles of the responsible person. Play the "what if" game—what will we do if the following happens? Once the action-reaction has been discussed, the person in trouble will know what to expect and is more likely to cooperate and help you rather than fighting you in blind panic.

Another consideration is how long it will take for rescue services to get to the accident site. This must govern your reactions. If the fire or ambulance station is two or three minutes away, leave it to the professionals. This is what they are trained for, and they have the tools and equipment. If you are a long way from help services, you will have to take action yourself. At the same time you should have known that you were far away and carried extra rescue equipment and possibly men with you.

Do not take an injured man to a hospital in your truck; in most cases an ambulance can be there within minutes. The ambulance crew are trained and carry the equipment they need. The time you think you might save could cause the injured person more trauma, shock, and discomfort.

Emergencies fall into two major areas:

- Emergency evacuation: everyone is mobile and able to exit unaided.

- Rescue: a member of the team cannot get himself out; he is either unconscious or incapable of movement under his own steam.

1. *Emergency evacuation*

Emergency evacuation of the hole or pipeline could be caused by any of the following:

- A change in the atmosphere, gas, or oxygen alarm

- Failure of a ventilation system in a sewer

- The minor injury or illness of a man who is an integral part of a team, serious claustrophobia panic

- A structural fault or disturbance which may cause structural failure

- External factors such as heavy rain clouds assembling in the area of your catchment. *Do not wait for it to rain.*

- Collapse or absence from his post of the responsible person or his designated standby person

- Collapse of bottom man or observer

- Unplanned or accidental discharge of industrial pollutants upstream

- Rapid and unexplained increase in fluid levels

- Over-heating

- Loss of lighting

- Loss of communication

An emergency evacuation is as a result of a hazardous or dangerous situation arising when no serious harm has yet taken place but the planned situation has changed. In this event, it is necessary to get all persons out of the confined space as quickly as possible without rush or panic. All persons can walk; some may need minor assistance.

The collapse of the observer could be considered a major or rescue emergency, but if standard procedures have been adhered to, it is not. The observer was the first man down; he unhooked the man lifter to allow others to come down. Once they were all down, he hooked himself back up to the man lifter. In any emergency, he is the first man out. He needs no assistance to hook himself up, as he is already hooked up. Gain the attention (by signals on lifelines) of workers in the space, and start them moving back to the hole. Hoist the observer out. By the time he is out, the others are in place, ready to be removed in turn.

In the event of a rapid increase of fluid levels, get the men moving back to the hole, hoist the observer out, and lower the man lifter. Get the lifelines of bottom workers, now in the hole, pulled up snug and tied off on the surface, and proceed with the removal of the crew in an orderly fashion. If there are step irons, do not wait for the man lifter; take up the slack on their lifelines and help them out, one at a time.

The same would apply to a gas alarm. You have a few minutes. Do not panic, but get them out as quickly and smoothly as you can. If there appears to be someone badly affected by gas, get him out first, after the observer. While doing so, phone for help by dialling 111 and ask for fire services. Your trained first-aider should attend to him until help arrives. If you have planned the job properly, help services should know where you are and what you are doing. They have the equipment necessary, and they will arrange the ambulance.

For the responsible person: In the event of an emergency evacuation, you are going to have a busy few minutes, but that is why *you* were chosen to be the responsible person. Keep a cool head, and your crew will be safely on the surface in no time. Rush it, and you will hurt someone.

2.   *Rescue evacuation*

Any of the following could require you to carry out a rescue. How you react to this kind of emergency can risk lives, cause loss of life, or save lives. Planning is all it takes. Stop, think a minute, and act as planned.

- An operator or person is injured, making him unable to walk or climb out of the pipe or manhole unaided

- Gas poisoning

- Oxygen deficiency

In the above cases, you must take great care in how you proceed. It is common for the recovery of injured persons to injure others. Confined-space rescue should, if possible, be left to the professionals in the fire service. If you were highly organised and had a man basket, you could possibly safely remove a man with a broken leg out of a pipe and manhole. This level of equipment is extremely rare and very expensive; the fire service has it, and you should use their services wherever practical.

Let's address each incident mentioned above:

a) *Serious harm injury, like a broken bone.* Assume that we did not have a good hold on an operator's lifeline and were not using the man lifter. The step irons broke away, and the man fell into the hole, breaking his hip on the batters. A bunch of assumptions can be made, what do we do?

We have two men on the surface, sufficient to pull him out. But any strain on the harness causes the man to scream in agony. Do we send one of the top men down to assist? Assist how? The one man on the surface alone could not pull the injured man out. Heaven help us if the second man was injured in some way too.

I would proceed as follows: I would maintain verbal contact to reassure the man in the hole and allow the other top man to call for help by dialling 111 and calling for the fire service. He would explain clearly what had happened, where we were, and how many people were involved. Once the rescue service was on the way, depending on circumstances, I might enter the hole (if I had enough persons to lift *me* out) to comfort the injured person while waiting for help. Upon arrival of help, I would again vacate the hole and allow the fire service to assess the situation and carry out the rescue.

b)   *Gas poisoning has caused the worker to collapse.* We will assume the simplest and most likely scenario to start with. We have tested, and the tester did not indicate gas present. We have an operator inside the manhole, pushing a drain hose or TV camera. He complains of nausea, and before we can get him out under his own steam, he becomes unconscious. If we have planned this job properly, the two men on the surface will pull the man out of the manhole using his lifeline or the man lifter, if used. NOTE: the man in the hole should *not* disconnect the man lifter block from his harness until he is back out of the hole.

Our trained first-aider will attend to him, making sure his airways are clear, providing resuscitation or CPR if required, and keeping him warm while the other top man phones 111 and asks for ambulance services, stating the location, the number of persons involved, and the suspected cause of collapse.

Confined Space Entry

c)   *A similar scenario to that discussed above.* In most cases, the injured person will recover fully once exposed to fresh air. You should still call the ambulance and let them check him out and possibly put him on oxygen-enriched air.

If either of these incidents were to happen up a pipeline, you would need to think carefully about what to do. We did not know there was gas there; he did not get out as soon as the alarm went off; and he has now collapsed. The observer can see him but does not have a gas tester. (He normally does not; it is carried by the man/men up the pipe.) The observer is not aware of the condition of the air around himself, let alone what has caused the collapse of the operator up the pipe. So what do we do?

1. We get the observer out of the pipe *now*, and leave the man in the pipe.

2. If we have the equipment, the standby rescue breathing apparatus is put on by the designated rescue person, and the rescue man enters the hole with the apparatus on and functioning.

3. The removed observer gets on the phone or radio and calls 111, asking for ambulance services, clearly indicating the location, the number of persons affected, and the suspected problem.

4. The rescue person pulls the unconscious worker out of the pipe towards the manhole, using the worker's lifeline. Topside persons can assist in this, but care must be taken not to snag or damage the lifeline.

5. Once in the manhole, the unconscious person is pulled up rapidly to the surface, where the first-aider takes over. If required, he will provide mouth-to-mouth resuscitation and CPR. He will continue until the ambulance arrives and the ambulance crew takes over.

6. The rescue person is removed as soon as possible after the injured person is removed.

Never try to affect a rescue of any kind unprepared. It is not uncommon for the person rescuing to be similarly overcome by fumes and for there to be two unconscious or dead men in the confined space. There are numerous recorded cases of this happening. It is hard to leave a friend up the pipe, but in the long run it is the best thing to do. It takes four times as long to get two men out of a hole than it does to get one out. Do it right the first time, and we will all live to tell the tale.

Ninety-nine per cent of emergencies are due to error; only 1 per cent are an act of God. Emergencies happen because we did not accurately identify our hazards. We did not plan for all eventualities. We did not play the "what if" game.

OTHER SAFETY INFORMATION FOR OPERATORS

Manual Lifting

1. All persons are to ensure the following:

    a) Hand and foot protection is worn.

    b) There are no obstructions in the direction of travel.

    c) The setting-down point is clear.

    d) You can see over the load when you are carrying it.

    e) You can lift the load; make a trial lift.

    f) If the load is too heavy or awkward, get someone to help you.

2. Getting ready to lift:

    a) Keep feet comfortably apart, one foot in front of the other, pointing in the direction you intend to go.

b) Bend the knees to a crouch position; your back should be straight, but not necessarily vertical.

c) Get a firm grip on the load with the palm of your hand; avoid fingertip grips.

3. Lifting and carrying

    a) With your chin pulled in and arms close to the body, lift with your thigh muscles by straightening your legs.

    b) Lift smoothly; do not jerk or strain.

    c) Lift by easy stages—that is from floor to knee to carrying position.

    d) Reverse the lifting procedure for setting down the load.

4. Dual lifting

    a) Lifting gangs must work as a team with one man giving instructions.

    b) Ensure the load is evenly distributed between the team.

5. Lifting aids

    a) Where possible, use lifting/moving aids, such as a hydraulic lifter or other approved manhole lifting equipment.

    b) Ensure you use them correctly.

BREATHING APPARATUS TERMINOLOGY

All breathing devices must be able to supply 40 to 44 litres of free air per minute per man.

BA gear	Breathing apparatus: any device or piece of equipment used to provide clean unpolluted air for the user
Respirator	A device that is fitted over the user's mouth and nose. Normally lung powered, using filter canisters to purify the air being breathed. Used to filter dirt or gasses from the air being breathed.
Positive pressure	Refers to the atmosphere inside a breathing air face mask. Indicates that there is a constant pressure inside the face mask greater than the external pressure. Upon breathing in the regulator, allows more air to pass until breathing stops. Normally about 1/2 bar. Prevents gas entering around face seal when breathing in. Recommended device.
Demand breathing device	No positive pressure. The user must breath in or suck air to activate the airflow. Typically has a SCUBA breathing mouthpiece.
Full-flow breathing device	Breathing air is constantly delivered to the face mask. Air will flow out through the discharge or around the face mask all the time. Uses lots of air.
Bail out set	A breathing air set normally incorporating a full face mask. Carried for use in the event of a gas alarm. Normally lasts for 10 minutes. Not normally used as breathing gear except in emergencies. Often supported by a surface supplied hose line and then used in the event the hose line supply fails. This is a recommended device for manholes.
Rescue BA set	A breathing set, normally of 30 minutes duration, totally self-contained with a backpack supporting a compressed air bottle supplying air to a full face mask, normally positive pressure
Full Face mask	Normally describes a full face mask incorporating a visor with a full-face surround that seals the face completely. Normally positive pressure.
Half mask	A mask that fits over the nose and mouth, normally supplied with air by a hose line attached to the user's belt incorporating a simple needle valve regulator. Most respirators are classified as half mask units but are fitted with filter canisters rather than an air hose.
SCUBA	Self-Contained Underwater Breathing Apparatus.
Breathing Air Bottles	Available as 10 minute, 30 minute, 45 minute duration. Duration varies depending on the amount of work being done. The above times are average.

Confined Space Entry

ENTRY CLASSIFICATION

CONDITION AS INDICATED ON SITE OR BY CLIENT	ENTRY CLASS 3	ENTRY CLASS 2	ENTRY CLASS 1	ENTRY CLASS 1 GAS OR PIPE ENTRY
STORM WATER LESS THAN 3 M DEEP	YES			
STORM WATER MORE THAN 3 M DEEP		YES		
SEWER LESS THAN 2.5 M DEEP	YES			
SEWER MORE THAN 2.5 M DEEP		YES		
STORM WATER MORE THAN 6 M DEEP			YES	
SEWER MORE THAN 5 M DEEP			YES	
PIPELINE 300 MM OR LESS	YES			
PIPELINE 310 MM OR MORE		YES		
GAS POSSIBLE (SEWER)		YES		
GAS PRESENT				YES
OXYGEN DEFICIENT				YES
HIGH FLOWS INDICATED			YES	
HIGH FLOWS POSSIBLE			YES	
CONTAMINATED				YES
DANGEROUS ACCESS			YES	
UNCONTROLLABLE ATMOSPHERE.				YES
MECH VENT REQUIRED TO MAINTAIN				YES
PIPE ENTRY				YES

MEN AND EQUIPMENT NEEDED FOR THE CLASS OF ENTRY	ENTRY CLASS 3	ENTRY CLASS 2	ENTRY CLASS 1	ENTRY CLASS 1 GAS OR PIPE
2 WORKERS	YES			
MAN LIFTER	PLUS	YES	YES	
3 WORKERS	OR YES	YES	YES	
4 WORKERS				YES
5 WORKERS				IF GAS
GAS TESTER	YES	YES	YES	YES
HARNESS AND LIFELINE	ONE	TWO	TWO	THREE +
EXTRA LIFELINE			YES	YES
EMERGENCY PIPE STRUT				YES
FIRST-AID KIT	YES	YES	YES	YES
HYGIENE KIT	YES	YES	YES	YES
BAILOUT BA		YES	YES	YES
FULL BA				YES
RESCUE BA			YES	YES
FIRST AID CERTIFICATED WORKER	IDEALLY	YES	YES	YES
RESPONSIBLE PERSON			YES	YES
RADIO OR PHONE COMMS	YES	YES	YES	YES
MECHANICAL VENTILATION		IF POSSIBLE	IF POSSIBLE	YES

Chapter Fourteen

High Pressure Water Injuries

FIRST AID

The first and most important point to remember is that all injuries due to high pressure water must be treated very seriously, and the worker must receive professional care as soon as possible.

The surface wound may vary from a long, ragged hole to a mere pinhole surrounded by an area of pale swelling. There may be considerable pain, or the wound may be numb. Since the surface wound does not in any way indicate the extent and nature of damage to underlying tissues, the first-aid treatment must be based on the assumption that injury to deeper tissues has occurred.

The surface wound should be dressed lightly to control bleeding and prevent further contamination. Where the limbs have been injured, it is best to support and immobilise them as for fractures. For injuries to the trunk the patient should always be placed in the foetal position (knees up towards the chest and on his side) and his airways kept open. Give nothing by mouth. Simple resuscitative measures and treatment of shock should be administered as needed.

The patient *must* be taken to hospital. Advise the hospital of clothing that may have been worn by the operator if such clothing may

High Pressure Water Jetting – An Operator's Manual

have been shredded and driven into the wound. If there was a glove or boot, send it with patient.

In the event that a person is injured by the impact of a water jet, the injury caused may appear insignificant and give little indication of the extent of the injury beneath the skin and damage to deeper tissues. However, large quantities of water may have punctured the skin, flesh, and organs through a very small hole that may not even bleed. At 100 litres per minute, that's 1.6 litres per second. In 1 second, you have 1.6 kilos of water inside you somewhere.

Immediate hospital attention is required, and medical staff must be informed of the cause of the injury. To ensure that this is not overlooked, all operators engaged on water jetting should carry an accessible, waterproof card which outlines the possible nature of the injury and bears the following text, or similar:

> This man has been involved with high pressure water jetting at pressures up to 2000 bar with a jet velocity in excess of 1,440 km/hr. Surface injuries sustained will not indicate the extent of injury to underlying tissue. Please take this into account when making your diagnosis. Unusual infections with microaerophilic organisms occurring at lower temperatures have been reported. These may be gram-negative pathogens, such as are found in sewerage. Bacterial swabs and blood cultures may therefore be helpful.

The cards mentioned above will be issued to you, and they must be carried in an easy to find place. If the injured person is unconscious, the card should be tied (through the hole provided) to his wrist or clothing. Ideally the other side of the card could have the operator's name and next of kin contact details, twenty-four-hour employer contact details, and so on.

High Pressure Water Injuries

Where surgical examination is not immediately possible in remote locations, first-aid measures should be confined to dressing the wound loosely and observing the patient closely until medical examination is available. The reason we do not dress the wound firmly is to allow water to run back out, if possible.

If any person is struck by the water jet, this fact *must* be immediately reported to a supervisor. Supervisors *must* send the injured operator to a hospital for treatment. Nerve damage is indicated by a loss of feeling and requires extremely urgent medical attention. If the wound is anywhere on the trunk, the patient must be sent to a hospital as a matter of extreme urgency, as internal organs may be punctured.

There is little point in sending your injured worker to an ordinary emergency outpatient hospital if no one there has any idea how to treat this type of wound. It is a specialist activity. Seek out the local injection injury specialist and ask permission to have his or her name and phone number printed on the emergency card. When sending anyone to hospital with this type of injury, phone the doctor before your injured person reaches hospital and tell him what has happened so he can take charge of the correct treatment.

I have almost lost my left hand because the emergency doctor refused to believe the extent of the injuries. Fortunately, my GP had a friend, and things eventually got sorted out. The wound looked like a large graze on my wrist, with no single puncture wound. All feeling had gone from the top of my hand, and my fingers felt as if I had placed them into boiling water.

A smart investment any operator can make is to do the St. John Ambulance advanced first-aid course. The life it saves may be yours.

THE NATURE OF THE INJURY
(Extract from a paper by Dr. K. Middleton, MB, ChB, MFCM)

The water jet has some of the qualities of a high-velocity missile in producing injury. The larger pumps deliver a force not unlike the impact of a .303 rifle bullet.

The entrance wound will tend to be small and give little indication of the extent and damage to deeper tissues. Along the track caused by the penetration of the jet, the surrounding tissues will be damaged if not killed by the shock waves developed as the energy of the jet is dispersed.

Deeper tissues will be lacerated, depending on the depth of penetration and the anatomy of the area. With abdominal wounds, there may be extensive laceration and damage to the viscera. Water will be forcefully injected into the tissues. It may be dispersed widely and will tend to follow the natural cleavage planes of the tissues. Where tissues are constricted, as in the hand or foot, considerable pressure may be exerted on nerve bundles and blood vessels, which may lead to further devitalisation.

Depending on the degree of the pollution of the water in the jet, injuries may be contaminated both by infection and foreign matter carried into the wound by the water.

Hospital Treatment
Treatment should follow the accepted principles for deep penetrating muscle wounds. The surface wound should be excised and the track of the jet carefully explored. Immediate wide surgical decompression should be carried out, especially in hand and foot injuries, followed by careful debridement along the length of the track. Contamination may require copious gentle irrigation and removal of debris by suction.

In abdominal wounds, all viscera must be carefully examined. Two cases have been described Neil and George 1969 and Dabrowski et al. 1994. In the first, the surface wound consisted of a semicircle of 5 cm radius comprised of pinpoint puncture wounds of the skin in the right iliac fossa, resembling very closely a superficial abrasion. In the second, there were two adjacent wounds on the left side of the abdomen 0.5 and 1 cm in size. In both cases, there were multiple lacerations and penetrations of the abdominal viscera.

Wounds should be packed open or closed loosely over drainage.

Chapter Fifteen

Personal Protective Equipment (PPE)

Nowadays *all* work requires some form of personal protective equipment. Safety glasses are essential in just about every endeavour, from car washing to micro surgery, from carpentry to water jetting. We have accepted safety belts in our car; we have accepted safety glasses and safety boots; and now we, the water jetting crew, need to go a little further along the PPE road and accept a bunch of other things that some of us resist and others gladly accept.

Personal protective equipment is not a luxury, and it is not intended to burden you unnecessarily. It costs a heap of money that management would rather not spend. Over the years, standards and codes of practice, insurance companies, and government departments have studied our accidents and injuries and have recommended certain levels of protective equipment to try to keep us safe and healthy.

Being humans, we have resisted protective equipment and have been injured as a result. It became obvious that legislation was needed to force us to use PPE when needed. I can remember back in the eighties when the union movement demanded reinstatement of an employee who refused to wear safety glasses. Unions now support employers over that one and no longer support an employee who abuses the PPE system. Times change.

Because of what we do, there are a number of things that can "bite" us. Flying debris can damage our eyes and exposed skin; noise can damage our hearing (not repairable); the water jet can cut us; chemicals and pollutants can make us ill; gasses can poison us; a lack of oxygen can kill us. We can fall off structures and into holes, and we can injure ourselves with many forms of sprain and strain. *All* of these can be minimized by using the appropriate PPE—using it all the time and using it properly. PPE can save your life.

A great deal of our PPE requires training to use. Difficult as it may be to understand, you should be trained even in the use of earplugs. Earplugs should be fitted to your ear; they are not a "one size fits all" device. The same applies to earmuffs and respirators; some fit my head better than others. Do you know how to check the quality of the lenses of safety glasses? Hold them about 200 mm off your eyes, and rock the glasses while looking through the lenses. Also bring them closer and move them away. The image you see must not distort, no matter what you do. This does not apply to prescription lenses.

If the client's PPE requirements are stricter than yours, go with the client's. If your in-house PPE requirements are stricter than the client's, go with yours. Never use less than your employer recommends.

Following are the minimum PPE requirements:

Sewer/pipe or vessel cleaning (non-entry)

- Waterproof gloves

- Wrap-around safety eyewear

Personal Protective Equipment (PPE)

- Safety boots
- Disposable coveralls
- Hearing protection
- Fall arrest if working over a hole

Sewer/pipe or vessel cleaning (entry)

- Waterproof gloves
- Monogoggles
- Safety Wellingtons with nonslip soles
- Wetsuit (oilies)
- Hardhat
- Hearing protection
- Harness and tether
- Man recovery winch after a certain depth (client regs)
- Gas testing device
- "HELP" calling device

NOTE: *All* manhole entry at any depth requires at least one other trained team member. A method of safely removing an unconscious man from the hole must be planned. See local client regulations.

Pipe cleaning

    As for non-entry sewer above

Working with a handgun

Gloves suitable for the task plus

- Face shield over safety specs or goggles
- Wet suit
- Safety Wellingtons
- Hearing protection

Carry in truck for all work
First-aid kit containing standard items plus:

  - Extra Elastoplast strips
  - Dettol or similar antiseptic wash
  - Optrex or similar eye wash and cup

- Injury report notebook

  Antiseptic hand soap

- Washing water for hand/face washing
- Towel, wipe rags, or disposable paper towel
- Drinking water and cup (allow at least 6 litres per day, more on very hot days and DRINK IT)
- Spare clothing, socks, and overalls, plus T-shirts, etc.

Personal Protective Equipment (PPE)

- An assortment of gloves

- Spare hard hat, safety glasses, and monogoggles

When working in sewers or waste pipes, you are not likely to pick up any diseases such as AIDS, hepatitis, etc. Many discussions have been held regarding inoculations against all manner of diseases, and there are almost no records of a disease being caught in a sewer. The amount of detergent, lavatory cleaner, and other household chemicals discharged into domestic drains makes the atmosphere most intolerant of viruses and germs (and you—gas test, gas test, gas test).

Some employers do inoculate their employees against disease, but this is normally a "duty of care" reaction and an effort to cover all the angles. If it makes you feel safer, go get it done.

You will get infections in cuts and scratches, however. It is therefore absolutely essential that all cuts and scratches are disinfected and covered with waterproof plasters as soon as they happen.

Always wash your hands and face carefully before lighting a cigarette or eating food. (*Do not light up till you are at least 10 metres from the open manhole.*) As above, you are not likely to pick up any diseases in a sewer, but a little care can go a long way in keeping you healthy.

The water jet can flush out angry bees, wasps, cats, rats, and snakes. Be aware of what may come out of the pipe.

Chapter Sixteen

Vacuum Loading

Let's discussion vacuum truck operation.

WHAT IS A VACUUM?
Atmospheric pressure acts on the surface of the earth at a pressure of one bar, 14.5 psi, or the equivalent of 10 metres of water head at sea level.

If we apply a vacuum to an area, we effectively remove the atmospheric pressure acting at that point, and loose matter moves up the column of vacuum. At the same time, the surrounding atmospheric pressure rushes to fill the gap and equalise the pressure. In this way we create an airflow up the vacuum pipe.

This airflow assists us in using vacuum as a tool and adds energy around the particles being lifted, aiding the movement of product. If we did not have this airflow, we would need a greater degree of vacuum to lift fluids. In other words, if we stuck the vacuum hose into the water, preventing the flow of air from entering the hose, we would need a far greater vacuum capacity to lift the solid column of fluid.

Let us look at this a bit deeper. Assume we had a sealed drum of water with a hose in the centre, buried deep in the water. This hose is open at the outside end. If we were to apply pressure to the drum, onto the surface of the water, what would happen? The water (pressure) would look for a way out. It would find that the pressure in the hose was less than the pressure in the drum, and the water (pressure) would escape up the hose and out to the atmosphere.

It follows that if the drum was open to atmosphere, and we reduced the pressure in the hose (by sucking), the pressure on the outside would try to equalise the pressure in the hose—and water would flow up the hose. Next time you stick a straw in a glass of coke, think about this.

If the hose was long enough and the suction great enough, the vacuum in the hose would lift a column of water until the height (weight) of the column equalled the pressure acting on the surface of the surrounding water. This is minus one atmosphere, or minus one bar, or 10 metres of suction head or 33 feet of suction head. At this point, we would have created a perfect vacuum. But this is not possible outside a laboratory.

A point to consider is that water is made up of two gases: hydrogen and oxygen. A combination of these two gases makes water when energy is applied. The pressure required to hold them together as water equals one atmosphere. If we could create a perfect vacuum and suck up minus one bar, or minus 14.5 psi, the molecules would separate into their respective gases.

SEPARATING THE SPOIL FROM THE AIRFLOW

One of the major problems with vacuum is preventing the dirt from flowing over and into the vacuum pump, filter bags, and cyclones. The concept of this separation process is airflow and speed. If the air is compressed inside a hose, it flows very quickly, thereby drawing up product and creating a localised vacuum. If you increase the size of the hose, the airflow is reduced, and it runs more slowly.

We have a blower (actually it sucks) that moves so many cubic feet of air per minute; this is fixed or constant. If the volume is drawn through a small space or restricted, the speed (cubic metres of air over x metres per minute) is great. If the space is suddenly increased, the speed (metres per minute) is greatly reduced. Solids suspended in a rapidly moving airflow travel fast and remain suspended in the airflow. Remember the laminar flow we talked about earlier? Air does exactly the same: the fastest air is in the middle of the pipe, and the solids run in that section; very little touches the walls of the hose. If you allow that airflow to enter a larger space (the bin), it will slow down rapidly. The solids previously in suspension will no longer be supported by the airflow and will fall out. Each stage of cleaning involves an increase in airflow speed by restricting it, followed by an increase in area, thereby reducing it again—each time dropping out more dirt.

Finally, the air is passed through a cyclone. The cyclone causes the air to spin at great speed. As the air spins, all heavy items are thrown to the outer wall, where they fall out of the airflow and settle in the catchment vessel below.

WORKING WITH A VACUUM
Under very special conditions in a laboratory, a perfect vacuum is possible. Mechanically, it is not possible. Outer space is a perfect vacuum. As there is no surrounding pressure to hold things together, water would turn to vapour and disappear. The very best we can get with any of our suction trucks, under ideal conditions, is minus 0.75 bar, or minus 11 psi. This works out to a closed suction lift of 7.5 metres as the absolute maximum. The average height of our suction vehicles is 3.6 metres to the ground, so it is safe to assume that pure suction of water or fluid from about four metres below the wheels would not be possible.

WORKING WITH AN AIRFLOW
However, the action of the vacuum blower (or sucker) draws 3,000 cubic feet of air per minute through an open-ended 6- or

8-inch pipe. This works out to an airflow of about 861 metres per minute or a wind speed of about 50 km per hour. Particles placed in front of the suction line restrict the airflow and are drawn, by the air flowing past them, up into the pipe and deposited in the storage drum. In this manner, suction heights can be greatly increased as the airflow can be through long lengths of hose. Suction heights of as much as 6, 8, or more metres can be carried out with ease *as long as you allow the air to be the main item travelling through the hose* or a larger percentage/volume of air than solids or fluid.

If, however, you stop the airflow by sticking the hose into a fluid or pile of muck, the unit's ability to lift will revert back to about 7.5 metres of lift from the product to the top of the bin. This is known as *working with a vacuum.*

It follows that with a permitted air leak into the end of the hose, the suction airflow of the blower is capable of far greater lifts than if you were to close off all air supply. This is known as *working with an airflow.*

Friction plays a major role too. Hose internals are rough, and friction on these roughened areas restricts the free flow of air in the same way friction restricts the free flow of water through a hose. By reducing the diameter of the hose, we increase the friction component and increase the effort needed to draw the air, speed, and flow through the line, thereby reducing the carrying capacity of the airflow. A 50 kph wind will blow over a bigger/stronger structure than will a 30 kph wind. This is a fairly basic comparison but an accurate one.

Let's take the friction component one step further: It follows that if the pipe is half full (what was a 6" hose is now a 3" hose) of moving product, the performance of the blower will eventually drop off. It is important to remember this and to allow the hose to take small portions at a time of the product being shifted. Allow each bite a few seconds to travel up the hose before taking another. Increases in friction reduce the airflow, and the moving product slows down until the pipe eventually blocks up.

Vacuum is normally measured in inches of mercury. This is directly related to the amount of vacuum (suck) you would need to provide to a tube filled with mercury. Sixteen inches would indicate that the unit can suck up 16" of mercury straight up like coke through a straw. One inch of mercury is 0.491 psi. Sixteen inches of mercury (the Guzzler) is 16 x 0.491 = 7.9 psi. Over an 8" pipe area, this equates to 4 x 4 x 3.142 x 0.491 = 196.6 kg. So if you had the hose pulling on your body, leg, or arm, it would exert a force on you of 397 lbs or 196.6 kg. Think about that the next time you stick your hand up the pipe to remove a blockage. Remember the advertisement on TV about some vacuum cleaner manufacturer claiming to be able to pick up a bowling ball? Well the Guzzler could pick up about sixty-five of them.

FIRE HAZARDS WHEN SUCKING

Fine particles travelling in an airflow rub together and cause friction which generates static electricity. As static builds up in a product, it transfers itself as an electric current to the metal structure in the form of mini lightning strikes. These flashes of electrical energy can ignite dust and cause a massive explosion. A major contractor in New South Wales was sucking fine wood dust from a furniture factory when the product ignited. The resulting explosion killed the operator, badly injured his assistant, and completely destroyed the Super Sucker. Always earth the machine to whatever steel structure there is nearby, and make sure that structure is itself earthed. If no suitable structure is available, you must provide your own: drive a steel or copper spike at least 600 mm into the ground; wet the surrounding area with water; and attach your earth clip to it.

Some products are more volatile—that is likely to ignite—than others, such as pine wood dust, sulphur, catalyst, coal dust, and milk powder. But to be safe, make it a standard practice to always earth the machine. This must be an instruction for all operators, and safe working practice must dictate *if dry suction, always earth*. It is a good idea to hook up a permanent, heavy-duty welding cable earth lead on a small reel on your truck and lay it out every time. Get into a habit that might save your life.

In some cases, the client may insist that your suction hoses are nonconductive and nonstatic. Great, but do yourself a favour and *earth* the truck too.

If wet suction is being carried out on fuel oil, crude oil, or other inflammable, wet product, prudence dictates that you fit an earth lead. The earth must be securely clamped to bare, unpainted steel at both ends. Ideally, an earth lead should be wrapped around the suction hose from end to end, and also run to earth. Be safe, not sorry.

I have seen flat-braided earth strapping, such as is used to earth your car engine to the chassis, self-tapper screwed to each suction hose joiner, taped along the outside of the hose, and eventually earthed. That way each metallic fitting was earthed to the next and then to the ground. I think it may be cheaper than nonconductive or anti-static hoses. Whether it's as effective, I am not sure.

It is good working practice to earth the vehicle to a tanker if refuelling from one vehicle to another. The same applies if you are transferring fuel from a drum on the back of one vehicle to the tank on another. Both vehicles have stored static electricity, and a spark can jump across via the fuel flow, igniting it. This is not farfetched; it has happened and is a serious hazard to be aware of.

Gases in sewers are flammable; a good spark could set them off. Unless you want a manhole lid around your neck, earth your truck.

Vacuum Loading

WORKING WITH A VACUUM—SUMMARY

We now know that there are two ways to use the vacuum system: as a direct suction unit or as an airflow carrying unit. Both systems use the same mechanics but work on different principles: either direct suction or airflow moving. Airflow moving gives a far greater lift capacity with restricted product flow rates, while direct suction gives a far greater product flow rate with a reduced suction head or height of lift.

Direct suction lift is rated not from the end of the hose but from the surface of the water into which the hose is placed. So if you wanted to suck sludge from the bottom of a tank filled with water, your limitations would be the height of the *top* of the water to the top of the vehicle inlet. The atmospheric pressure acting on the surface of the water around the outside of the hose will push the water up inside the hose to an equal level; from there you must suck it.

At the same time, the weight of the product being lifted must also be considered. The weight of a column of water is 1 kg per litre and requires a good amount of energy to move it. Sludge out of a drain could weigh as much as 2 tonnes per cubic metre, or 2 kg/litre. The weight of milk powder is 0.2 kg per litre and will require far less energy to move. Powders are grains of product which are in themselves light, small, and normally spherical. They travel well in an airflow. But fluids are heavier than air and fall, during travel, to the lower half of the pipe and try to hold on to the pipe wall using friction. The air rushing over the surface of the lying water draws the surface of the water along, but the water against the pipe may be stationary. As the water fills the area of the pipe, vacuum increases but airflow reduces, and the two methods of travel or lift change as indicated above.

It follows then that if "air travel" is being used, you should take great care to locate your suction hose in such a way as to prevent dips that may fill up with water. You need a continuous slope with all dips and hollows eliminated. Humps are acceptable.

TRAVELING ON THE ROAD—VEHICLE LOADING

Almost all vacuum units are capable of loading well in excess of their permissible road load weights. Remember that 1 cubic metre of fluid weighs at least 1 tonne (some solids weigh even more). Most vehicles are currently licensed to carry nominal loads on the road, not a full bin. When you need to carry a full bin on a public road, you need to get a temporary permit to do so. The tare of the vehicle (its dry weight) is detailed somewhere on the cab on the driver's side. Add the weight per metre of the product you intend to carry, and work out your load capacity this way.

If you do not know the weight per metre of a product, you may find it detailed in a list at the end of this chapter. Some allowance can be made for air space around the product if lumpy, but you should stay on the side of safety and use these numbers only as a guide.

TRAVELING ON THE ROAD—HAZARDOUS PRODUCTS

Before you can drive a truck carrying hazardous substances on a public road, you have to attended an Approved Driver Training Course, Hazardous Substances. Your driving license will be endorsed HAZARD CERT with the date in the bottom right-hand corner. If you do not have it, you are breaking the law. This does not apply on private property, although it is recommended.

Many of the products we are asked to suck are noxious (harmful substances to health or the environment), poisonous, corrosive, or explosive. Once we have them in the bin, the customer is no longer responsible for them; we are.

For example, a well-known company in Australia was asked by another very well-known international company to remove a solvent from a tank and get rid of it. They sucked up the product into their Super Sucker without any particular precautions and proceeded to the disposal site. Upon arrival, they were asked by the person who was to receive it for disposal what the product was.

"Oh, just a solvent", the driver said. "But what solvent? What is in it?" they asked. "God alone knows", said the driver. The outcome was that the disposal people refused to take it until they had a full and accurate description of the product. It turned out that it contained PCBs, and no one wanted it. It had to be sent to France for disposal. The contractor went back to the client and asked him to take it back. But the client had quickly demolished the tank and was not the least bit interested in taking it back. He felt that once the contractor had accepted the contract, it was his responsibility to dispose of it.

After several days of argument and discussion, during which time the Super Sucker sat in the contractor's yard, the contractor and the client came to an agreement to split the costs. Special plastic-lined drums were purchased, and the product was transferred under very strict conditions—full-body chemical suits, breathing apparatus, standby fire department, ambulances, etc.—into the drums.

Meanwhile, there was a veritable war going on in France with the Greens, who were demonstrating against their country being used to dispose of other country's waste. The facility was shut down.

The contractor now had twelve drums of something awful and no place to get rid of them. A concrete tank was built—complete with bund walls, fences, and security systems—into which were placed the twelve drums of a "simple solvent". I suspect they are there still.

The moral of this story is, Never accept any product unless you know exactly what it is *and* you have arranged with the disposal company to take it. In most cases, the client should do this. At the same time, he should fill in a Hazardous Substances Dangerous

Goods Declaration form. This form is the only document you may accept to transport dangerous goods. Do not accept any other.

It is good working practice to collect this form before you pick up the product. It tells you

- its proper shipping name and common name;

- Its hazard class;

- its UN number (international recognition);

- its flash point (the temperature at which it will catch fire or explode—the lower the temperature the more dangerous it is to suck and to handle; adding vacuum may reduce the flash point even further);

- shipper, the name of the company sending the goods;

- consignee, where it is going to be dumped;

- carrier, you; and

- an emergency contact phone number so that responsible persons can phone the owner of the product and get details.

If the form is fully (all sections) and correctly filled out, you and your company are fully covered.

This form is carried on a clipboard or pocket on the driver's door inside the cab of the truck carrying the product. If you have an accident, the rescuers, fire department, or whoever can find it and attend to the spillage in the correct and proper manner.

The box marked "HAZARD CLASS" indicates which of the labels you *must* display on the back and front of your vehicle. There are nine basic with twenty-one subgroups. If you have a load of mixed product, you should indicate both. If the product being carried has more than one hazard, all signs should be displayed.

It follows that it is good practice to check what hazardous goods you may be handling prior to leaving the yard. This way you can plan ahead and make sure you have the correct vehicle labels. Ask ops or your rep to get the declaration faxed through when he gets the order to work, so you can examine it.

Some products are so dangerous that you should wear a chemical suit and breathing apparatus when handling them. The protection provided to you—such as disposable overalls, gloves, a facemask or respirator, and goggles—are to protect you against the product whose toxicity is known. There is little point in wearing all the prescribed clothing to suck it up and to wear nothing when you discharge it out of sight of the client.

Protective clothing is for the purpose as detailed at *all* stages of the handling process. When cleaning bags, *find out what has been sucked, and take precautions as for that product*. No one should ever clean bags without full PPE, a respirator, and protective eyewear, preferably goggles.

MIXING HAZARDOUS SUBSTANCES

There are many chemicals and substances that may explode, give off poisonous gas, or react violently if mixed one with another. These are not always obvious. *If in doubt, don't.*

If the information available to you is not clear, call your safety officer, who has more detailed reference books.

It follows that it is good practice to wash out your bin, cyclones, and hoppers each time you discharge product *at the discharge site*, not in the yard. It is noxious; that's why the client asked you to handle it.

Some common products to be wary of:

- Caustic soda

- Peroxide, either organic or hydrogen (do not mix with anything)

- Any oxidant or oxidising agent

- Chlorine

- Any products from tanneries

- All acids

- Anything ending in *ic*, *ate*, *yde*, or *ide* is an indication that it may react dangerously with something else.

- Any product marked "Dangerous When Wet", as there is often water in the hose or bin. There is often water where you discharge the load.

- A chart detailing the separation distance between chemicals is available from your local Work Safe or Work Cover Department. Get one, and keep it in your truck.

WARNING
Do not ever breathe fumes coming from the vacuum pump discharge. They could kill you.

VEHICLE OVERLOAD
It is very easy to overload your truck. Most will carry 7 or 8 cubic metres. If this is wet sand or something similar, you are carrying a load of 16 tonnes plus your tare. The traffic police do not like that all that much.

Vacuum Loading

NOTE: Most of the figures in the following table are given for a solid block or hard, compacted lump of material. You will not suck a solid block nor will you compact it in your bin. However, be aware of material weights and how they may affect you in an overload situation.

MATERIAL	WEIGHT/CUBIC METRE
Water (1,000 litres)	1,000 kg or 1 tonne
Most oils and crude (1,000 litres)	900 kg
Aluminium solid	2,560 kg
Brass or bronze solid	8,000 kg
Borax	1,800 kg
Brick	1,800 kg
Calcium	1,570 kg
Cast iron solid	7,200 kg
Chalk	400 kg
Cement	3,100 kg
Coal	1,500 kg
Concrete	2,200 kg
Glass	2,600 kg
Granite	2,650 kg
Gravel (generally)	1,750 kg

Iron sands (loose, steel mill)	2,800 kg
Lead	11,137 kg
Limestone	1,850 kg
Masonry (general)	2,700 kg
Plaster of Paris	1,800 kg
Quartz	1,800 kg
Salt	2,100 kg
Sand, dry	1,600 kg
Sand, wet	2,000 kg
Sandstone	2,300 kg
Sewage (as water)	1,000 kg
Soil, common black	2,000 kg
Steel solid	7,800 kg
Sulphur	2,000 kg
Tar	1,000 kg
Wet, compacted wood chip	680 kg

TIM'S USEFUL NUMBERS/CONVERSIONS

1 Atmosphere (Atm) (bar)	=	14.5 psi (14.73)
1 Atm (1 bar)	=	1 bar (1.013)
1 Atm (1 bar)	=	10 metres vertical head
1 Atm (1 bar)	=	33 feet of vertical water head
1 Atm (1 bar)	=	1.02 kg/cm^2
1 Atm (1 bar)	=	100 kPa
-1 Atm (-1 bar)	=	30" mercury vacuum
1 psi	=	6.9 kPa
1 MPa	=	10 bar
10 kPa	=	1.45 psi, or 1/10 bar
1 lb/in^2	=	0.07030697 kg/cm^2
1 kg/cm^2	=	14.22334 lbs/in^2
1 kg/m	=	6.72 lbs/ft
1 Newton (N)	=	0.22480697 lbs/in^2 force
1 inch^2	=	6.4516 cm^2
1 m^2	=	10.7639 ft^2

High Pressure Water Jetting - An Operator's Manual

1 litre of water	=	1 kg
1 litre	=	0.264 US gal
1 litre	=	0.222 imperial gal
1,000 litres	=	1 tonne (metric ton)
1,000 litres water	=	1 cubic metre of
1,000 litres	=	264.2 imperial gal
1 tonne	=	1,000 kg or 2204.62 lbs
1 metre	=	39.37 inches
1 inch (")	=	25.4 millimetres
1 foot (')	=	304.6 millimetres
1 kilowatt (kW)	=	1.4 horsepower
1 horsepower (hp)(PS)	=	0.746 kilowatt
1 inch of mercury	=	0.03386388 bar or 0.4911541 psi
1 inch of mercury	=	13.58 inches of water head
convert degrees C to F	=	(C x 1.8) + 32
convert degrees F to C	=	(F—32) x 0.56
area of a rectangle	=	length x width = m^2

Vacuum Loading

area of a circle	=	radius x radius x 3.143 = m²
circumference of a circle	=	diameter x 3.143 = m
internal volume of a vessel	=	area of the base x height = m³
volume of a cone	=	area of the big end x half the height = m³
area of the wall of a round tank	=	circumference x height = m²
Hose pressure rating	=	BP to be at least 2.5 x MWP
Hose testing press. & period	=	6 monthly to 1.5 x MWP for 5 minutes.
Male and female screwed fittings	=	min. 6 full turns
Thread sealant	=	Loctite 567 to 10K and Loctite 667 10K+
Bursting disc/Rel. Valve setting	=	20% above MWP of weakest component
SWP sure	=	Safe Working Pressure
MWP	=	Maximum Working Pressure
BP	=	Burst Pressure
TP	=	Test Pressure (normally 1.5 x MWP)

Minimum Bend Radius of hose	=	Normally 24 x OD
Reaction force (R=Kg)	=	R = 0.0227 x ltrs/min x $\sqrt{(bar)}$
Water Velocity (V= m/sec) Cd = 1	=	V = ltr/min ÷ (bore mm) d^2 x 21.22
Coefficient of Discharge (Cd)	=	Measured discharge ÷ theoretical
Water jet velocity	=	V = 14.14 x $\sqrt{P}$ = M/sec
Best velocity of water in hose to eliminate friction	=	Should not exceed 18 m/sec

meters per second

Water kW	=	bar x ltrs/min x 0.002 = kW
Water horsepower	=	bar x ltrs/min x 0.002 ÷ 0.746 = hp
Engine power required	=	Water kW or hp. x 1.6
Hose	=	Always measured inside diameter (ID)

-4 + 1/4"; -8 = 1/2"; -12 = 3/4"; -16 = 1"

metric hose is described in mm id

Vacuum Loading

Pipe = Always measured inside diameter (ID)

1/2" ID pipe has OD of about 3/4"

Tube = Always measured outside diameter (OD)

1/2" tube ID—2 x wall thickness

Some Calculations (all pressures in bar)

*kW = $\dfrac{\text{l/min x pressure x 2}}{1{,}000}$ = maximum continuous power kW

*l/m = $\dfrac{\text{kW x 1,000}}{\text{pressure x 2}}$ = flow from a pump

*bar = $\dfrac{\text{kW x 1,000}}{\text{l/min x 2}}$ = max operating pressure

*Plunger velocity = $\dfrac{\text{Stroke x 2 x rpm}}{60 \times 1{,}000}$ = metre/sec

*Flow = $\dfrac{\text{Piston dia}^2 \text{ x 0.785 x stroke x no. of plungers x rpm}}{1{,}030{,}000}$

= Litres per minute

*Copied from Hammelmann chart Lf/me.

TUBE & PIPE CLEANING

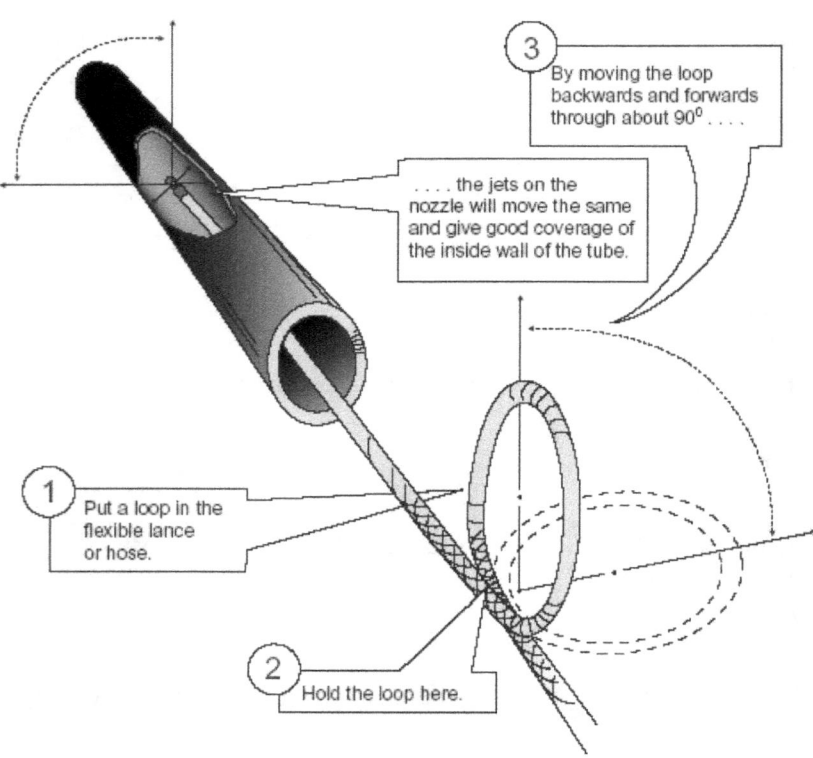

Vacuum Loading

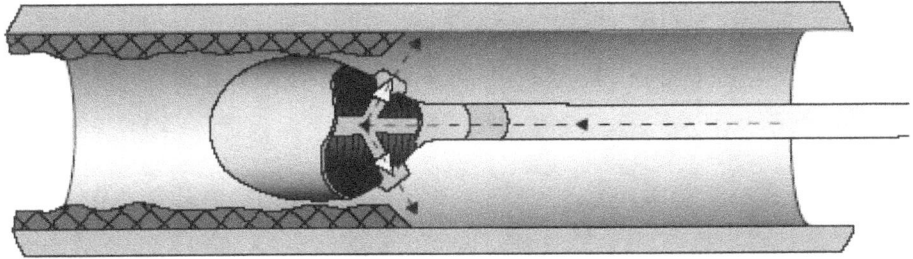

Where the pipe is large in diameter and the deposit tough - NOZZLE CARRIERS can be used.

These are large nozzles into which screw-in nozzles are fitted.

This brings the nozzle closer to the deposit.

As the jet is separately made to the carrier, a better orifice profile can be made.

Giving a more efficient jet.

PROBLEMS TO WATCH FOR WHEN CLEANING TUBES

It may be necessary to clean heavily blocked tubes progressively when using flexible lances or hose BECAUSE:

> Debris being brought back will pack around the flexible lance and the walls of the tube.

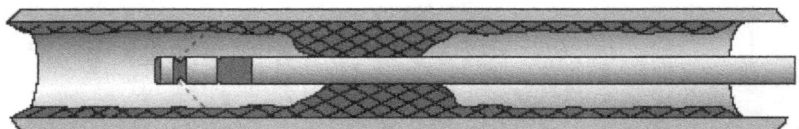

THEREFORE

> If the outside diameter of the flexible lance or hose permits little clearance between itself and the inside bore of the tube and / or the build up of deposit in the tube is heavy.

CLEAN PROGRESSIVELY

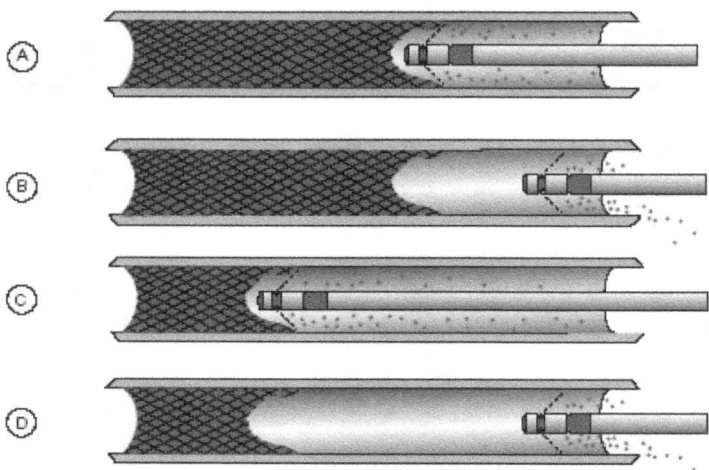

Normal tube cleaning with flexible lances or hose backward facing radial jets at 45⁰ - giving good thrust to propel the nozzle up the tube and flush back any deposit removed.

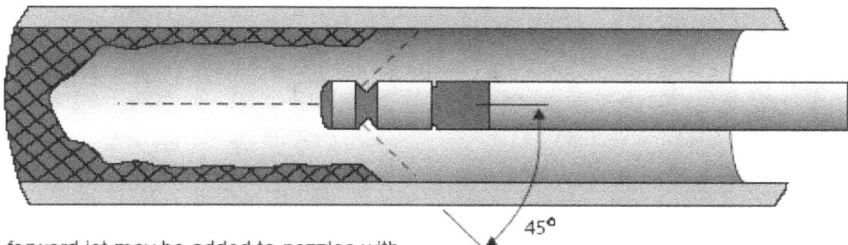

A forward jet may be added to nozzles with
radial jets at 45⁰ to cater for light blockages or where build-up of deposit prevents passage of jet by reducing bore below outside diameter of jet.

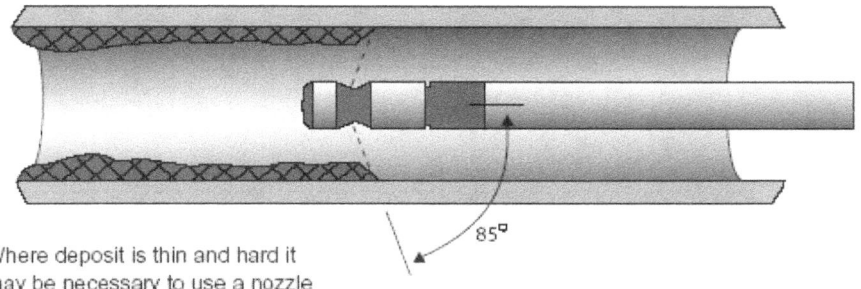

Where deposit is thin and hard it
may be necessary to use a nozzle
where the radial jets hit the internal surface of the tube more directly. The jet is angled back, just sufficient enough to provide forward thrust.

Where tubes have blockages of a very hard or stubborn deposit, forward facing jets on ridged lances may be the only solution.

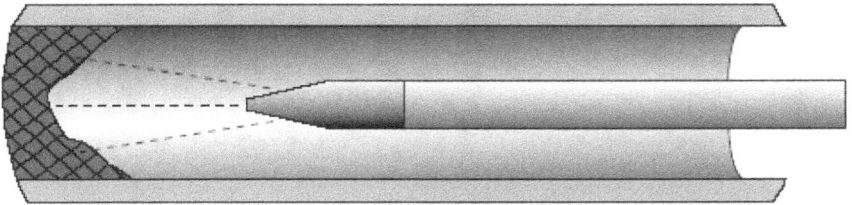

When using a rigid lance whose outside diameter comes close to the wall of the tube - thus allowing little room for deposit to pass (or water to pass), a 'piston like' effect will force the nozzle out of the tube.

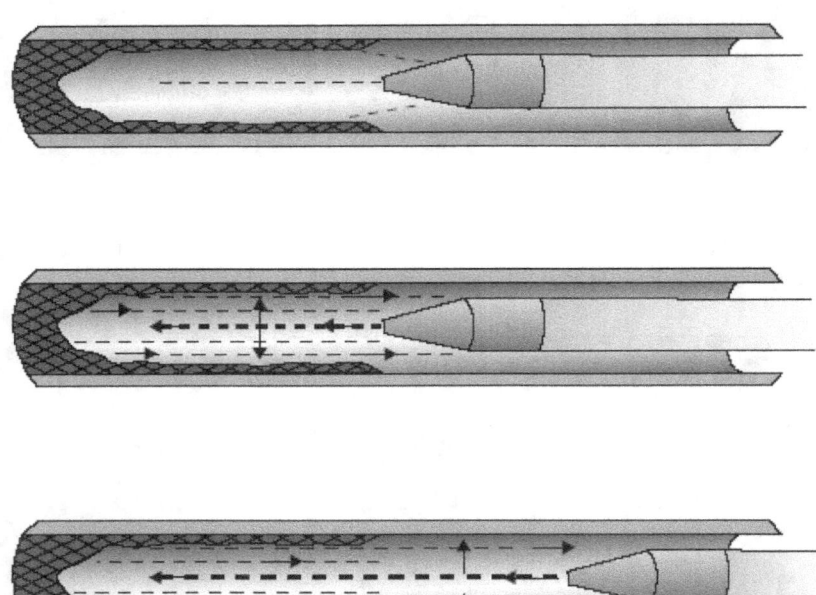

Although this may be obvious in blocked tubes. Remember that it may be possible to create a blockage in an unblocked tube by pushing removed debris forward.

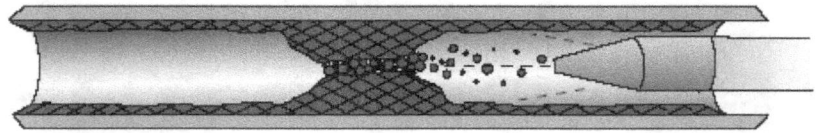

Vacuum Loading

Normally radial jets are 4, 5, 6, or 7 in number, giving reasonable diametral cover.

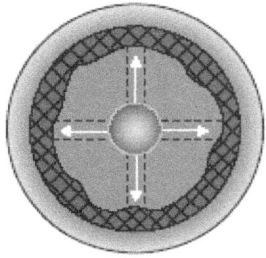

Where deposit is very tough it may be necessary, apart from changing to 85° jets, to concentrate as much power as possible into the jet.

This can only be done by reducing the number of jets and enlarging their size.

Such jets with only 2 or 3 radial jets are usually found only in the 85°B nozzle.

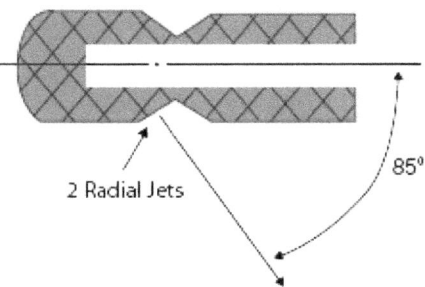

2 Radial Jets

85°

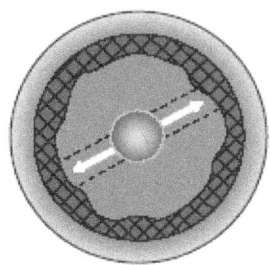

*NB - It will be necessary to rotate the nozzle through at least 180°.

www.ingramcontent.com/pod-product-compliance
Lightning Source LLC
Chambersburg PA
CBHW071520180526
45171CB00002B/321